CANADA IN SPACE

The People & Stories behind Canada's Role in the Exploration of Space

Chris Gainor

The Publisher: Folklore Publishing
Website: www.folklorepublishing.com

Library and Archives Canada Cataloguing in Publication

Gainor, Chris
   Canada in Space / Chris Gainor

Includes bibliographical references.

ISBN 1-894864-59-X (pbk.)

       1. Outer space—Exploration—Canada—History. 2. Astronautics—Canada—History. I. Title.

TL789.8.C3G34 2006   629.4'0971   C2006-905803-2

Project Director: Faye Boer
Project Editor: Bridget Stirling
Production: Patricia Begley, HR Media
Cover Image: courtesy of NASA
Photography credits: Every effort has been made to accurately credit the sources of photographs. Any errors or omissions should be directed to the publisher for changes in future editions. *Photographs courtesy of* the Canadian Space Agency (p.26, p.30, p.38, p.45, p.73, p.119, p.124, p.128, p.133, p.144, p.148, p.154, p.168, p.179, p.181, p.183, p.193, p.197, p.200, p.202, p.208, p.225, p.233, p.242, p.247); NASA (p.81, p.88, p.97, p.109, p.111, p.190, p.204, p.213, p.217).

We acknowledge the financial support of the Alberta Foundation for the Arts for our publishing program.

We acknowledge the financial support of the Government of Canada through the Book Publishing Industry Development Program for our publishing activities.

 Canadian Heritage Patrimoine canadien

Contents

~ာ၄၉C၈~

Dedication

To Taylor and Brian, Aidan and Sarah, Amanda and Taylor, and Lyra and Solomia, who will witness our future endeavours in space and maybe even live them.

Acknowledgements

Thank you to Denise Thomson, Richard Bobier, Mary Gainor, Bill Woytenko, Mark Gainor, Kirby O'Connor, Sandra McClellan, Norman and Kathy McClellan, Dr. Robert W. Smith, Dr. Stephen B. Johnson, Peter Davis-Imhof, Barry Shanko, Ken Harman, Andrew Yee, Kieran Carroll, Randy Attwood, Paul Fjeld, Frank H. Winter, Katherine Zwicker, Brian Gold, David Lee, Tom Sheppard, Gwen Walter, Tom Hawthorn, Patrick Nagle, Rolf Maurer, Steven Pacholuk, and my parents, Don and Toni Gainor. Special thanks to my wife, Audrey McClellan, without whom this book would never have been written.

Introduction

EXPLORING SPACE IS BOTH A SIMPLE THING THAT ALMOST everyone can do and one of the most challenging endeavors humans can undertake. Outer space is easy to reach and accessible because the skies are open to anyone who goes out in clear weather and looks up, but it's highly challenging for those who want to journey beyond the Earth's atmosphere.

This book looks at the Canadians who have opened the roads into space with rockets and satellites. But to understand what drew them to the stars, we must look back at those people who explored space with their feet firmly planted on the ground.

For most of recorded history, the only way to explore space was to look up. Most people today live in well-lit homes in light-polluted cities with televisions, computers and activities to keep them busy. But most of these products are the innovations of the past century. Before that time, humans had dark skies to look at after sundown and little to distract them. Urban Canadians today can only see the star-filled skies when they get away from cities.

Those who do look up into the night sky see thousands of stars of varying brightness, and some of those stars make up a band known as the Milky Way. They move in the sky not only in a daily cycle but also in an annual cycle that has stars visible in at different times of the night according to the seasons.

Ancient peoples used these annual cycles to set their calendars. Some other things that can be seen with the naked eye in dark skies are fuzzy objects that are actually galaxies and nebulae. Along with the Moon, a handful of lights now known as planets move against a relatively fixed background of stars. Until 400 years ago, no one had telescopes to help look more closely at the skies and the objects in them, so people wondered about what they saw when they gazed up at the night sky.

Recorded observations of the sky go back almost as long as any civilization known to historians. The ancient Egyptians and Babylonians made observations and kept records of the movements of the Sun, Moon, planets and stars thousands of years ago. They kept these records to make calendars and predict the arrival of seasons for agricultural pursuits. Astrologers looking for omens in the sky also used these records. Early Chinese and Indian astronomers made similar observations. Neolithic peoples in England and elsewhere built structures like Stonehenge, which appear to have been used as temples and observatories to predict when the Sun would reach the northernmost and southernmost points of its annual movements in the sky. Inspired by the Egyptians and Babylonians, the ancient Greeks and Romans developed our astronomical knowledge, followed by Arab astronomers during Europe's Dark Ages and then by European scientists during the Renaissance.

In recent years, archeologists and historians have come to recognize that the cultures of the aboriginal peoples of the Americas are as rich as the cultures of the peoples who came from across the oceans to populate what today is known as Canada. Today, we know that that the aboriginal peoples were the first to look up into the skies above Canada and try to make sense of what they saw. For thousands of years, they have lived in North America's varying landscapes and climates, from the Arctic—where the Inuit seasons vary from the midnight sun of summer to darkness at noon in winter—to the fertile lands more than halfway to the equator in what is now southern Canada. While most Native peoples are thought of as hunters and gatherers, only some made their living this way. Others got by as fishers, and still others were engaged in agriculture. Many Native peoples needed to keep track of the seasons, and to do that, they looked to the sky for the changing signposts of the seasons that people in other parts of the world also followed.

The Native people of North America used celestial observations to help them determine seasons, and more importantly, they also used them to plan the rituals that enriched their lives. The Tsimshian and other peoples along the Pacific lived by fishing and travelling up and down the coast, so their lives required knowledge of the sky to help them navigate and anticipate events such as salmon migrations. The Tsimshian used the regular cycles of the Moon, which were used for calendars throughout the ancient world, to indicate the seasons and the availability of food. These coastal peoples developed complex societies and called upon their elders to use their knowledge of the Sun and other celestial objects to help make sense of the natural world. Some West Coast peoples looked for omens when they journeyed into the sky during dreams. Stars, Sun, Moon and winds were seen as humanoid immortals, along with other sky dwellers of legend.

Shamans of the Ojibwa and other Algonkian peoples who live in today's Ontario speak of their origin as the descendents of people who once lived on the Moon and came to the Earth after passing through what is usually translated as a hole in the sky. Their migration to the Earth came when a drought struck the Moon and was accomplished with the help of a gigantic spider. Ojibwa shamans regard the Pleiades star cluster, visible on winter evenings, as the opening where spiritual power passes between this world and the star world. The shamans also speak of a comet that burned the ground around Lake Superior, leaving it to people to fertilize.

"To the Ojibwa, the knowledge preserved in the origin tale can be seen and interpreted in the surrounding landscape," according to Thor Conway, who gathered information from the shamans. "At night, they see their history in the stars." Today, most of us are just beginning to learn of the rich variety of native cultures in Canada, and part of that knowledge is that which they gathered from the skies for centuries before Europeans crossed the oceans.

The first Europeans to come to what is now Canada needed to be informed about the night sky to make it here. Starting with John Cabot's voyage to New-foundland in 1497, just five years after Columbus came to North America, those who crossed the Atlantic did so with the help of primitive navigation instruments such as quadrants, astrolabes and cross staffs that allowed navigators to determine the angle between an object such as the Sun or Polaris (the North Star) and the horizon. By sighting Polaris at night or the Sun at local noon, the navigator could determine what lati-tude he was at, giving him his position relative to the equator and the North Pole. Determining longitude (one's location east or west) was difficult in the days before the chronometer was invented in the 18th cen-tury, making it possible to tell time at sea. The highly difficult methods of determining longitude in the early days involved infrequent astronomical events, such as eclipses or the passage of the Moon in front of stars, which take place at known times. In those times, cloudy days made navigators' work difficult or impos-sible for a time.

When the first European explorers came to Canada, the telescope had not been invented, and most people still believed that the Sun, the Moon and the planets orbited around the Earth. Only the theoretical work of Copernicus in the 15th century and Galileo's telescopic observations of the planets and the Moon starting in 1609 led knowledgeable people to start believing that the Earth and the other planets orbited the Sun. Just as importantly, Galileo's observations showed that the Moon had craters and mountains and that the planets were more than points of light in the sky—they were worlds unto themselves. Soon people began to think about what it might be like to visit these places, although most ideas were fanciful. Many writers imagined that these worlds were inhabited, and some, such as Cyrano de Bergerac, imagined flying into space using bottles of dew that were raised into the sky by the rising sun or soaring into the heavens with the help of a fanciful flying machine.

While Europeans grappled with these startling ideas of new worlds in the sky, the first settlers in the real New World on the other side of the Atlantic were dealing with the sometimes-hostile environment that they found. One of these environments was the first settlement in New France at today's site of Québec City. These courageous pioneers focused on survival, learning about their new home and finding a basis for a new economy. Just as the sailing explorers needed to know the sky to navigate, so did the people who were surveying the new lands and trading routes.

Other sky watchers, such as the Jesuit fathers, came to New France starting in 1604 to establish missions among the aboriginal people. The Jesuits were also known for their educational institutions, which trained missionaries who combined scientific knowledge and curiosity along with their religious work. Father Paul Le Jeune observed eclipses of the moon from the colony in 1633 and 1635. He records explaining lunar eclipses to a puzzled Native man. Other Jesuits used their ability to foretell eclipses to prove the superiority of their beliefs. Jesuit missionaries also recorded astronomical observations and worked to establish the latitude and longitude of Québec and other settlements. Civilian authorities were engaged in similar activities. Jean Bourdon, who was named surveyor general of New France in 1634, used his knowledge of astronomy to do his survey work. In 1646, the Jesuits gave him a small refracting telescope to help him with this work. Jean Deshayes, who had trained at the Paris observatory, arrived in New France in 1686 and soon used observations of a lunar eclipse to determine the longitude of the colony at Québec.

In the 1750s, Britain and France began fighting over their possessions in North America after skirmishes that had begun years before. The war culminated at the Plains of Abraham just outside Québec City, in 1759, when British forces led by General James Wolfe defeated the French under Montcalm and ushered in the period of British rule over what today is Canada. Although the departure of the French meant the end of the Jesuits' educational efforts in the area, the incoming British surveyors, sailors and explorers were also trained in astronomy and used this knowledge to map their new territories.

In 1769, the Royal Society in London sent an expedition to Fort Churchill on Hudson Bay to observe a rare transit of Venus across the face of the Sun. Such transits were important to astronomers because they gave them a means of calculating the distance between the Earth and the Sun. Captain James Cook, who took part in the British conquest of Québec and later made a celebrated expedition to the South Pacific to observe the 1769 transit of Venus, used his astronomical knowledge in 1778 to fix the location of Nootka Sound on Vancouver Island. His former midshipman George Vancouver did the same when he returned in 1792 to the coast of today's British Columbia to survey the area.

Much of Canada was being explored by men from the Hudson's Bay Company, which had been set up by an English royal charter in 1670. Samuel Hearne headed west from Hudson Bay in 1769 and mapped and explored much of the land to the west and north up to the Coppermine River and its route to the Arctic Ocean. Because he used a primitive quadrant, many of Hearne's latitude measurements were inaccurate. But many other Hudson's Bay explorers were more adept at their measurements thanks to efforts of company surveyor and teacher Philip Turnor. One of the most notable explorers was David Thompson, who produced accurate maps of nearly one-fifth of the continent in his work for the Hudson's Bay Company and then for the North West Company, the fur trade competitor he joined in 1797. His work included mapping the boundary between the British and American territories west of Lake Superior, much of the area around the Rocky Mountains and the entire length of the Columbia River in today's BC and northern U.S. Alexander Mackenzie, a colleague of both Turnor and Thompson, put his astronomical knowledge to use during his historic overland trip to the Pacific Ocean in 1792–93.

Today we know that the Earth is surrounded by a gigantic magnetic field and that the lines of Earth's magnetic field touch the Earth at the North and South Magnetic Poles. These poles are not the same as the Geographic North and South Poles, and the magnetic poles change position. This fact is important to Canada, because the North Magnetic Pole has been located in Canada's North throughout recorded history, and it is to this pole that compasses point. The history of exploration in Canada's North and of Canadian astronomy are strongly related to our possession of the North Magnetic Pole. The British Royal Navy dispatched expeditions to Canada's Arctic in the 19th century to search for the Northwest Passage to Asia and for the North Magnetic Pole, which was crucial to navigators.

These explorations, which were supported by the prestigious scientific group called the Royal Society, carried scientists who made geomagnetic observations and recorded their impressions of the geology and life, both human and animal, in the Arctic. These expeditions included John Ross' expedition of 1818 and William Edward Parry's expeditions of 1819–20, 1821–23 and 1824–25. All these adventurers failed to find the Northwest Passage, but they made observations about changes in the Earth's magnetic field. Parry also tried to reach the North Pole by ship in 1827, but he fell well short. Ross returned for an expedition starting in 1829, and despite the fact that his ship was trapped in the ice and his crew was stranded for four years, he lost few men before his

rescue. He found the location of the North Magnetic
Pole in 1831 on the Boothia Peninsula. In 1845, Sir
John Franklin famously led a similar expedition that
ended in tragedy after his two ships were trapped in
the Arctic and he and his men died. Other expeditions
were sent in search of Franklin and his crew and
continue the quest for the Northwest Passage. At the
same time, the stars and other celestial bodies
became something more than just targets for navigators—they became objects worthy of study by
scientists in Canada.

CHAPTER ONE

Astronomy in Canada

MANY PEOPLE DATE THE FORMAL BEGINNING OF ASTRONOMY IN Canada to 1839, when the British government decided to establish the Toronto Magnetic and Meteorological Observatory along with magnetic observatories in other parts of the world. In the wake of the expeditions to Canada's North, the Royal Society and the Royal Navy were engaged in what became known as the "Magnetic Crusade." Their goal was to understand changes in the Earth's magnetic field that affected compass readings—no small matter for the world's largest navy. Changes in the magnetic field were then believed to affect weather as well. The observatories were set up under the supervision of British Army Major Edward Sabine, a well-known scientist who was a veteran of the War of 1812 in Canada and of the Ross and Parry Arctic expeditions.

The Toronto observatory was built in 1840 on what is part of the University of Toronto today. The original structure was made of logs and included a small telescope. A permanent stone structure for the observatory was erected on the site in 1855.

Sabine's findings from Toronto and other observatories led him to conclude that storms on the

Sun directly affected the Earth's magnetic field. Among other things, he found that the Earth's 11-year geo-magnetic cycle directly followed the Sun's 11-year sunspot cycle.

The presence of the Magnetic North Pole in Canada brought further expeditions to the North when scientists from 11 nations took part in a major geo-physical research effort, the International Polar Year of 1882–83. An American expedition to Lady Franklin Bay on Ellesmere Island in northern Canada became what American science writer Walter Sullivan called "one of the most terrible episodes in the history of the Arctic." Manned by personnel from the U.S. Army Signal Corps under the command of Lieutenant Adolphus W. Greely, the group of 25 inexperienced polar explorers set up their base near Lady Franklin Bay in August 1881, a year before the start of the Inter-national Polar Year. The following spring, a sledging party came closer to the North Pole than anyone before when it explored northern coast of Greenland. But the anticipated relief ship failed to show at the end of summer, and so the group waited another year, using up their food, before beginning to make their way south, The journey was harrowing and deadly. By the time a ship found the group in June 1884, only eight members of the party were alive, and two of them didn't survive the remainder of the journey home. A German group also went to Baffin Island, and Canadians made observations at Fort Rae on Great Slave Lake as part of the Polar Year explorations.

The scientists' efforts resulted in new information about the Earth's magnetic field. To build on that knowledge, 50 years later, scientists from 44 countries took part in the Second International Polar Year, in 1932–33. This time, scientific interest was increased because of the discovery that waves from radios, which had been invented since the First International Polar Year, were bouncing off a layer of charged particles high in Earth's atmosphere, and these waves were affected by Earth's magnetic field. Canadian scientists set up stations to record data on weather and magnetic changes inside the Arctic Circle and used radio-equipped balloons to gather information.

By the time of the Second International Polar Year, the Canadian government had a body that carried out scientific activities. The group got its start as an advisory committee set up in 1916, in the middle of World War I. In 1925, the committee became known as the National Research Council of Canada (NRC). The NRC established laboratories to work on various scientific issues in the early 1930s, including physics. The work of these laboratories grew during World War II as physicists grappled with problems such as radio communication, which was critical to the war effort.

Canadian scientists were also looking beyond the Earth's atmosphere and its magnetic field. In the remaining years of the 19th century, some small observatories were set up for keeping time and occasionally for astronomical observations. But the 20th century opened new vistas for astronomy and science in Canada, starting with the construction of the Dominion Observatory in Ottawa at the beginning of the century. While astronomy in the U.S. developed with the help of wealthy patrons, Canada's progress followed a different path. Richard Jarrell, in his history of astronomy in Canada, wrote that: "A striking feature of the growth of Canadian society and its economy, and equally of the sciences, is the significant role of the state." And so it was with Canadian astronomy. The federal government built the new observatory, which began with the basic work of timekeeping along with research in astrophysics and solar astronomy.

An astronomer at the Dominion Observatory, John Stanley Plaskett, worked to persuade the federal government that a larger facility was needed. He soon succeeded in having the Dominion Astrophysical Observatory built on Little Saanich Mountain, just north of Victoria, BC. When the 1.8-metre telescope began operation in 1918, it was the largest in the world, until a bigger telescope in California was completed a few months later. Plaskett, who served as director of the observatory until 1935, discovered many binary (double) stars, including a large one named in his honour. Along with Joseph A. Pearce, Plaskett made important findings about the rotation and structure of the Milky Way galaxy. They found that our home galaxy rotates once every 220 million years, and that the Earth is located about a third of the way from the edge of the Milky Way.

Astronomers need to be educated, and the University of Toronto astronomy department, set up under the leadership of C.A. Chant, trained many of Canada's early astronomers. In 1935, the university gained importance in astronomical circles when the David Dunlap Observatory was established in Richmond Hill, just north of Toronto. The telescope is slightly larger than the Victoria telescope, making it the largest in Canada, and it was built thanks to a rare example of scientific philanthropy, when the family of Toronto financier David Dunlap donated money for the observatory. Many leading Canadian astronomers learned their craft at the University of Toronto. During this time, astronomers such as Frank Hogg and Helen Sawyer Hogg also assumed positions of leadership as professional astronomy in Canada began to grow after World War II.

In the 20th century, many people began seriously to think about travelling into space themselves. For hundreds of years, people had fantasized about space travel, and these dreams were made more vivid by 19th century novelists such as Jules Verne. In the first three decades of the 20th century, theorists such as Konstantin Tsiolkovsky in Russia, Hermann Oberth in Germany and Robert Goddard in the United States began to show that the way into space was the rocket. Until then, rockets made of gunpowder were used as fireworks or as military weapons of limited utility. But these theorists and others showed that rockets with powerful new liquid fuels could potentially carry payloads and people high into Earth's atmosphere and beyond. Goddard successfully launched the first liquid-fuelled rocket in 1926.

Publicity about the theories of these space pioneers caused hundreds of people in Russia, Germany, the United States, England and elsewhere to form rocket societies and promote space travel. Many of the people who joined these groups would go on to take leading positions in the German rocket program of World War II and the Russian and American space programs that began after the war.

Canada had a modest group of its own. In 1936, a 15-year-old German refugee named Kurt Stehling got 20 people to join a rocket club that he set up at Central Technical School in Toronto. The group built a few gunpowder rockets and tried its hand at designing spacecraft, but the club disbanded when World War II began, as did most of the other rocket societies in Europe and the U.S. Stehling and others kept their enthusiasm for space exploration alive during the war, and Canada would take a leading role in the first wave of space travel after the war ended in 1945.

CHAPTER TWO

John Chapman and the Beginning of Canada's Space Program

CANADA STRETCHES FROM THE NORTH POLE MORE THAN halfway to the equator, and its most southernly point lies farther south than parts of California. As the crow flies, the distance between Victoria, BC, and St. John's, Newfoundland, is more than 5000 kilometres. In total landmass, Canada is second only to Russia. And although most Canadians live in a narrow band along the southern border, there are settlements in every part of the country. So, since the creation of Canada in 1867 and its expansion in the years that followed, keeping in touch has always been a big priority and a big problem for the Canadian government. In Canada's early years, its government championed the building of the railroad. The telegraph and the telephone soon followed, but they connected only the populated parts of southern Canada.

The 20th century brought radio. Guglielmo Marconi received the first wireless message to cross the Atlantic in St. John's, Newfoundland, in 1901, and in the 1920s, the use of radio and other wireless communications exploded. Radio allowed communities from every part of Canada to keep in touch, but a problem remained. Radio communication in the North was not always reliable.

The unreliability of radio in the North is related to the fact that radio communication over long distances is possible because radio waves bounce off the ionosphere, a dynamic region in the atmosphere starting 60 kilometres above the Earth's surface and extending up thousands of kilometres into the near-vacuum of space. The ionosphere gets its name because the atoms there gain and lose electrons, and hence become ionized. The ionosphere is affected by solar storms, which unleash charged particles that strike Earth's magnetic field and the particles in the ionosphere. These storms affect radio communications, cutting off radio contact with northern settlements for days at a time. Because the Earth's North Magnetic Pole lies in Canada's North, Canada feels the results of solar storms and their effects on the ionosphere more than any other country on Earth.

Another effect of these ionospheric interactions is the northern lights (aurora borealis) that light up the night skies in northern Canada and even, on occasion, in southern parts of Canada. Today, tourists visit the North for a view of the dazzling auroral displays.

The scientific exploration of the Canadian Arctic by British and American expeditions in the 18th and 19th centuries led to further scientific research into the magnetic field and the ionosphere. The invention of radio gave new relevance and prominence to this research.

Scientists from many countries came to Canada to conduct geophysical research during a major coordinated research effort known as the Second International Polar Year in 1932–33. As a result of that work, the National Research Council of Canada, Canada's major civilian government research agency, began conducting ionospheric research in the 1930s. When radio communication assumed greater importance during World War II, the NRC increased its research in this area. NRC scientists working under physicist Frank T. Davies measured the ionosphere during the war in order to better understand how it affected radio communications, especially naval communications. The federal government set up the Defence Research Board (DRB) in 1947. Its branch, the Defence Research Telecommunications Establishment (DRTE) in Shirley's Bay, near Ottawa, took up the wartime research on the ionosphere's impact on military communications and radar. The DRTE, which was also run by Davies, included the radio propagation laboratory, which carried out much of the DRB's ionospheric research, and the electronics laboratory, which concentrated on improving radio and other communications hardware.

Scientists bounced radio waves off the bottom of the ionosphere to record its changing reflective properties. They launched balloons and sounding rockets to probe the upper atmosphere. American and Canadian scientists began launching sounding rockets in 1956 at Churchill, Manitoba, on the shores of Hudson Bay. The site was ideal for the study of aurora borealis and the ionosphere close to the Magnetic North Pole. At first, American rockets were launched, and the U.S. military operated the site, but the facility came under Canadian control in the 1960s. Canadian-built Black Brant rockets carried most of the research payloads launched at Churchill after that. Sounding rocket launches continued until the Churchill site was closed down as a result of 1980s government budget cuts.

Although the sounding rockets and the balloons gave brief glimpses of the ionosphere, Earth's magnetic field and the causes of auroras, the scientists wanted to learn more. In the 1950s, space researchers began planning for the first artificial satellites to be launched into orbit around the Earth.

The push for satellites came to a head during the follow-up research effort to the two International Polar Years known as the International Geophysical Year (IGY), which took place in 1957 and 1958. During the IGY, 60,000 scientists from 66 countries worked to increase their knowledge of Earth's atmosphere and magnetic field by taking measurements in every part of the Earth, including at the two poles, underneath the oceans and high in the atmosphere. The IGY is best remembered for inspiring the launch of the first satellites, and it was the successes of the IGY that inspired scientists to look at satellites as a way to build on their knowledge of the ionosphere.

When the Soviet Union launched *Sputnik* into orbit on October 4, 1957, the date was marked as the beginning of the space age. Less than a month later, the Russians launched a dog, Laika, into orbit aboard *Sputnik 2*. Early in 1958, the U.S. launched its first satellite, *Explorer 1*, which discovered the first evidence of radiation belts surrounding the Earth. During 1958, a number of American and Russian satellites found the full extent of the radiation belts. In the United States, a new agency called the National Aeronautics and Space Administration (NASA) was formed to explore space. One of NASA's missions was to foster international scientific cooperation.

The work of the IGY and the discoveries from the early satellites inspired a group of Canadian ionospheric scientists to propose a satellite that could probe the ionosphere from above, bouncing radio waves from the topside of the ionosphere. Late in 1958, Canada's Defence Research Board (DRB) proposed building this satellite to NASA. The following year, NASA and the Canadian government agreed that NASA would supply a launch vehicle, and Canada would design and build the satellite, *Alouette 1*, at the DRB under the direction of a tall, bespectacled, 38-year-old Canadian physicist named John Herbert Chapman.

Chapman was born in London, Ontario. After serving as a radar officer in England during the war, he studied physics at the University of Western Ontario and McGill University in Montréal and then joined the staff at the Defence Research Telecommunications Establishment (DRTE). Chapman's PhD thesis at McGill was on ionospheric radar echoes, and much of

Dr. John H. Chapman

~⚙~

his research in the 1950s dealt with ionospheric physics. At Western, he met a fellow student, Marian Holmes. They married in 1949 and had five children.

Chapman was promoted to section leader of the ionospheric propagation unit at the DRTE in 1951, and he soon became a leading researcher there. Chapman's expertise in ionospheric physics and his leadership qualities led to his appointment as head of the Alouette program when it began in 1958.

Phil Lapp, an engineer at De Havilland Canada who played an important role in Alouette and other Canadian space efforts, said that Chapman: "seemed to have his own private agenda. And a guy who was very hard to penetrate—he kept things pretty close to his chest. But a person who I swear had a vision about what was going to happen in space long before anybody else did. And I think that's what makes him a real pioneer."

"Chapman was a natural leader," said physicist Doris Jelly, who worked for Chapman in the early 1950s. "Even though Chapman was always in a hurry, he had a twinkle in his eye and a spark of fun...I remember the time we found him demonstrating his version of the twist dance craze."

"Chapman, was a very quick and intelligent person, with little time for small talk," former colleague LeRoy Nelms recalled. Chapman worked with another DRB scientist, Eldon Warren, to develop the idea of Alouette, he said.

Chapman's team was made up of many experts, including Keith Brown, who headed the engineering team; Colin Franklin, who was put in charge of electrical systems; and John Mar, who headed mechanical design. Alouette would need long antennas to do its work, and the team found the design they needed in an antenna called the STEM (storable tubular extendible member). The antenna was rolled up like a carpenter's measuring tape during launch, and when it was unspooled in space, it would form a tube and maintain its strength as it extended. The antenna was the brainchild of one of Canada's greatest inventors, George Johnn Klein, whose inventions included nuclear reactors and electric-powered wheelchairs.

The Alouette team and the scientists working for the NRC and the Department of National Defence get credit for beginning Canada's space program. But there were other institutions, such as the University of Toronto Institute for Aerospace Studies, that also contributed to fostering Canada's first generation of space pioneers.

With de Havilland Canada and Avro Canada building leading-edge aircraft such as the Avro Arrow in the 1950s, many young engineers working on those programs began thinking about flying beyond the atmosphere. Some of those engineers formed the Canadian Astronautical Society, and their colleagues working for Montréal-based aircraft contractors such as Canadair formed the Astronautical Society of Canada. The Toronto group began working on a rocket called Charm. Members from the two groups were recruited for Alouette. In 1962, both groups joined with other associations of aeronautical engineers to form the Canadian Aeronautics and Space Institute, which continues to support aeronautics and space activities in Canada today.

As for the defence scientists at Shirley's Bay, they were fascinated by Sputnik even before they had the Alouette satellite program. When the Soviet satellite was launched, Shirley's Bay scientists joined in the race to observe the satellite passing overhead, and they were one of the first groups in the Western Hemisphere to correctly calculate its orbit.

The equipment to be flown on Alouette, including the STEM antenna, was tested on sounding rockets, and in June 1960, the U.S. launched a Transit navigation satellite that included the first Canadian equipment sent into orbit, preparing the way for Alouette by testing for background radio noise in space.

Alouette 1 was designed to carry ionospheric sounding experiment and radiation detectors to add to the trove of information about the Earth's radiation belts. The Canadian team took a conservative approach to building its new satellite, which proved useful late in the process when NASA informed Canadian authorities that the vibrations during launch would be more severe than originally anticipated. *Alouette 1* was already strong enough to face the greater buffeting it would experience in flight.

"When we started the program, we knew virtually nothing about the problems of designing and building a satellite," Franklin recalled. "There were no textbooks on the subject. You had to write your own textbook as you went along."

"Most people on the Alouette team were young and worked long hours—70 and 80 hours a week," he said.

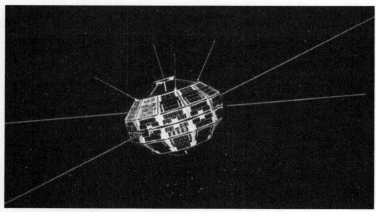

Alouette 1, Canada's first satellite

~∞∞~

"We tested very thoroughly all the bits and pieces on the satellite." At least one small electrical part gave Franklin "queasy feelings," but the part passed all its testing, so Franklin cleared it for flight.

For a variety of reasons, the original 1962 launch date was postponed for several months. Three weeks before launch, Franklin remembered, a Canadian magazine predicted that Alouette's launch vehicle would fail or the satellite would go dead if the launch succeeded.

Finally, two flight-ready 145-kilogram oval-shaped Alouettes, each covered with 6500 solar cells, were flown to Vandenberg Air Force Base near Lompoc, California, where the U.S. launched satellites into polar orbits. One of them was installed inside the nose cone on top of a Thor-Agena rocket. Chapman later admitted he was nervous as the launch approached. "I had my fingers crossed, my legs crossed, and everything else crossed. At that time, there was a 50-percent chance of failure in launchings."

Late on the evening of September 28, 1962, the countdown reached its end and the Thor-Agena rocket soared south into the black skies over the Pacific. A few minutes later, *Alouette 1* was in orbit, 1000 kilometres above the Earth. For those left behind in Shirley's Bay who listened to the countdown in the Defence Research Board facility's cafeteria, it was already September 29. An hour and a half after launch, *Alouette 1* neared the end of its first circuit around the Earth and passed over a station in Alaska that picked up the signal confirming that it was in orbit. The signal meant that the Canadians could celebrate. Canada's first satellite had not only gone into orbit, but Canada had become only the third country—after the U.S. and Russia—to have a satellite it had built put into orbit.

Prime Minister John Diefenbaker was quick to offer congratulations: "The knowledge gained, which will be used for peaceful purposes, should greatly improve the problem of communication. The launching of the Alouette with the United States supplying the rocket power is a practical illustration of the cooperation for peaceful purposes which exist between the United States and Canada."

At the time *Alouette 1* was launched, most satellites had short lifespans. *Alouette 1* was designed to last for a year, but its builders decided that if it lasted three months, the mission would be deemed a success. Ten years after launch, after it had sent back two million soundings of the ionosphere, *Alouette 1* was turned off for the last time after setting a record for length of operation in space. More than 280 scientific papers were generated from *Alouette 1*'s data alone, also a record for the time, and in them scientists shared their findings about how the ionosphere reacts to changes in the solar wind. By any measure, Canada's first satellite was a remarkable success.

"The *Alouette 1* was probably also the most complicated satellite that had been built up to that time," said LeRoy Nelms, who worked in the program.

"Colin [Franklin] may have been the biggest single factor in our success. He used our in-house scanning electron microscope to perform [quality assurance] on our supplier's electronic components and decided that commercial transistors were of too low quality for space applications.

"We ended up paying the manufacturer to set up a dedicated production line in order to produce transistors, which met Colin's quality standards. This may have been the birth of the space components industry. It was certainly the reason that the *Alouette 1* lasted 10 years, rather than the industry standard lifetime of several weeks that we had experienced up to then."

Adding to the Canadian pride was the fact that a simpler American ionospheric research satellite that was scheduled to fly before *Alouette 1* was held up by technical problems and didn't fly for another two years. Soon after *Alouette 1*'s success, Canada and the U.S. signed an agreement setting up the International Satellites for Ionospheric Studies program (ISIS). Both countries agreed to launch satellites to continue probing the secrets of the ionosphere, and scientists from other countries also took part in the ISIS program.

Canada's first contribution to ISIS was *Alouette 2*, which was launched from Vandenberg Air Force Base on November 29, 1965. Although the satellite was originally the backup for *Alouette 1* and carried similar experiments, it marked an important step in the creation of a space industry for Canada. The first Alouette was built by the Defence Research Board with de Havilland supplying the antennas and other equipment. The job of rebuilding *Alouette 2* was contracted out, with RCA Canada Limited of Montréal and de Havilland refurbishing the satellite and upgrading its equipment. One of the major movers behind this decision was Chapman, who would continue to dedicate a great deal of effort in the years to come to encourage the growth of Canada's space industry. *Alouette 2* was launched into an orbit nearly 3000 kilometres high together with an American satellite that conducted companion experiments in a nearby orbit. The second Canadian satellite lasted almost 10 years and nearly equalled the scientific productivity of *Alouette 1*.

RCA was also prime contractor for Canada's two ISIS satellites, along with the special projects and research division of de Havilland, which had worked on the two Alouettes. In 1968, the division was spun off from de Havilland and became Spar Aerospace.

Both ISIS satellites carried more sophisticated equipment to probe the ionosphere as well as tape recorders to increase the amount and variety of data that could be sent to Earth. *ISIS 1* was launched from Vandenberg on January 30, 1969. *ISIS 2*, which carried equipment to image auroral displays from above, was launched on March 31, 1971. Both spacecraft operated longer than the two Alouettes and were handed over in 1984 to Japanese scientists. But the Canadian government shelved plans to build another ISIS satellite as its priorities shifted to communications. More than a generation passed before Canada built another scientific research satellite.

John Chapman had moved on to other endeavours for most of the ISIS program's time, so David Florida managed the ISIS program until his death just before the launch of *ISIS 2*. Canada's major spacecraft assembly, integration and testing facility, completed at Shirley's Bay in 1972, was named the David Florida Laboratory in his honour.

Even before the second Alouette satellite was launched, major developments were shaking the world of communications and satellites. Various communications satellites, such as *Echo* and *Telstar*, had been launched into low Earth orbits, but they provided unsatisfactory results since they were visible from ground stations for only short periods of time. In 1963, all of that changed. NASA launched *Syncom 2* into an orbit roughly 36,000 kilometres above the equator. At that high altitude, the satellite remained stationary in relation to the ground below. With special equipment on board the satellite, it could receive radio, television and telephone signals beamed from Earth, amplify them and send them to another location on Earth thousands of miles away. The next year, *Syncom 3* beamed live television from the 1964 Tokyo Olympics to audiences in the U.S. and Canada. In 1965, a more powerful version of Syncom, *Early Bird*, was placed in a similar orbit above the Atlantic Ocean, which was the beginning of regular satellite television and telephone services across the Atlantic.

The implications of these satellites weren't lost on the Canadian government. Similar satellites could mean reliable television, telephone and radio service for every part of Canada. The success of the Alouette and ISIS programs and the growing potential of space technology in general meant that it was time for the government to think about where Canada could and should go in space.

In 1966, the federal government set up a four-member group chaired by John Chapman to study space and upper atmospheric programs. The other members of the group were Phil Lapp of de Havilland Canada, who had also worked on Alouette; Gordon Patterson, the founder of the University of Toronto's Institute for Aerospace Studies; and P. A. Forsyth from the University of Western Ontario.

The study group held hearings across Canada in the summer and fall of 1966. It heard from Canada's aerospace industry, government agencies and universities. In February 1967, the group completed a 258-page report, *Upper Atmosphere and Space Programs in Canada*, which surveyed the current state of those programs and laid out what most people believe was the blueprint for Canada's space program.

The report, which has been known ever since as the "Chapman Report," recommended the establishment of a Canadian space agency, the launch of Canadian communications satellites, consideration of Canadian satellites for remote sensing of resources on Earth, the development of a Canadian launch vehicle for small satellites in low orbits and the establishment of a Canadian space industry that could export satellites and related products to other countries. The Chapman report was silent on the possibility of Canadian astronauts, probably because the expensive and dangerous race between the U.S. and Russia to land humans on the Moon was in full swing. There appeared to be little prospect of other countries sending their own astronauts into space at the time.

The recommendations from Chapman's group had various outcomes. The Canadian government acted to establish a domestic communications satellite system and to support Canadian industry in the space field, but it did not set up a Canadian space agency until more than 20 years later, in 1989. Canada took no steps to develop satellite launch vehicles, but the country is presently an active participant in the U.S.

human spaceflight program, supplying hardware and astronauts.

When the Chapman report came out, Canada was in the middle of a wave of patriotic fervour as it celebrated its centennial. Chapman's summary of his work fit in perfectly with the mood of that time: "In the second century of Confederation, the fabric of Canadian society will be held together by strands in space just as strongly as railway and telegraphy held together the scattered provinces in the last century."

The Canadian government was most interested in what Chapman and others were saying about communications satellites. The government was anxious to provide broadcast services to all Canadians in both official languages, and the federal cabinet was so eager to get a communications satellite that it considered buying *Early Bird*, which was already in orbit. Instead, it decided that Canada should build its own satellite.

While these discussions went on in 1968, Canada elected a new prime minister, Pierre Elliott Trudeau, who reorganized the government and decided to set up a Department of Communications. To head it, he appointed a dynamic politician, academic and businessman from Montréal, Eric Keirans. The new minister hired Chapman to help him set up the department, appointed him assistant deputy minister and gave him responsibility for communications satellites. In 1969, the government created a new corporation to build and operate Canada's new communications satellites, Telesat Canada.

Dr. John H. Chapman at the launch pad in California shortly before the launch of *Alouette 1*

While Chapman's call for communications satellites was heeded, the government did nothing about developing Canadian launch vehicles. Instead of setting up a Canadian space agency, the federal cabinet set up an Interdepartmental Committee on Space and named Chapman as its chair.

At the beginning of the 1970s, Chapman was busy helping his department and Telesat Canada with communications satellites. He also pursued his goal, endorsed by the government, of building Canada's space industry. Hopes were high that RCA Canada could build Canada's first Anik communications satellites, but the government reluctantly chose an American builder, Hughes Aircraft, which offered to build a satellite far sooner and for less money than RCA Canada. The government stipulated that Hughes use Canadian components, so Hughes began calling on companies such as Spar Aerospace to build components for all its communications satellites. The RCA Canada aerospace plant in Montréal became part of Spar Aerospace, making Spar Canada the dominant Canadian space company in the 1970s.

Chapman still continued to pursue the dream of Canadian-built communications satellites through government programs that put Canada at the leading edge of satellite development, most notably through the Communications Satellite Technology program. The satellite built under the program, *Hermes*, pioneered direct-to-home broadcasting and other satellite technologies.

Larry Clarke, Spar's founding CEO, credited Chapman for the growth of his and other Canadian firms. "Dr. Chapman, as the federal government's representative, stood in the wings as a sort of godfather ensuring that Canadian industry was ready, willing and able to stand up to the challenges as they evolved. Above all, Dr. Chapman was determined that the Canadian industry must establish itself on a competitive basis so as to be able to expand from the domestic sales base into the international market."

As head of the government's committee on space, Chapman received an unexpected invitation from NASA in 1969 for Canada to join its human space programs after the Apollo Moon landings. At first, Canada chose not to take part, but when NASA decided in 1972 to build the space shuttle and again asked for Canadian help, Chapman was among those who encouraged the government to build the robot arm for the shuttle that became known as the Canadarm.

In 1979, Chapman was busy pursuing his goals for the Canadian space industry. That year, Telesat announced that Spar had won the competition to build the *Anik D* communications satellite, realizing Chapman's dream of Canadian-built communications satellites. A new government took office and launched a review of Canadian space programs that many hoped would lead to a Canadian space agency. In August, Chapman led a joint government–industry mission to Australia that demonstrated Canada's capabilities in the satellite field. But on September 28, the 17th anniversary of the launch of *Alouette 1*, John Chapman was in Vancouver preparing to give a speech when he died suddenly of a heart attack.

"Canada has lost an extraordinary individual," Communications Minister David MacDonald said. "Dr. Chapman played a major role in virtually every space activity in Canada. Canada's space program is where it is today to a very large extent because of his activities.

"One of Dr. Chapman's current enthusiasms was the idea of providing direct-to-home TV by satellite to people in remote and rural areas of Canada," the minister said.

One of his former colleagues said, "Chapman would have liked to have become the first head of a Canadian space agency, but unfortunately he didn't live to see that happen."

On Chapman's desk back in Ottawa on the day he died was a letter from NASA inviting Canada to send its own astronauts into space aboard the space shuttle. The invitation was finally taken up nearly four years later, once the Canadarm proved itself onboard the shuttle in 1981.

But Chapman's dream of a Canadian space agency would wait nearly another 10 years after his death before the government of Prime Minister Brian Mulroney set it up in 1989. In his life, Chapman won many awards from engineering and scientific organizations, but the honour that might have pleased him the most was bestowed on him in 1993, when the Canadian Space Agency opened its new headquarters in St. Hubert, Québec, just south of Montréal. The building was named the John H. Chapman Space Centre. The agency's most prestigious award is also named after Chapman. Many people who worked alongside him to build Canada's space industry have won the honour.

Thanks to John Chapman's vision, Canada became an important participant in world space activities with a strong space industry. The direct-to-home satellite broadcasting that Chapman championed has become a part of everyday life for people in Canada and around the world.

CHAPTER THREE

Canada's Rocket Scientist

THE TERM ROCKET SCIENTIST HAS COME TO BE SYNONYMOUS with people who build rockets and spacecraft, although most of them are actually engineers. And Canada has become famous for building satellites and robot arms, not rockets. Yet Canada has made its own memorable contributions to rocketry. One was a rocket that built up a unique record of achievement but is virtually unknown to most Canadians.

The story begins in the aftermath of World War II at the beginning of the Cold War between the Western countries, led by the United States and its allies, including Canada, and the Soviet Union and its communist satellite states. One of the Canadian government's responses was to set up the Defence Research Board in 1947, and one of the DRB's largest installations was the Canadian Armament Research and Development Establishment (CARDE). CARDE was established at the end of World War II at the large military reservation at Valcartier, just north of Québec City. One of CARDE's early projects was developing a Canadian air-to-air missile known as Velvet Glove that could be installed on aircraft such as the Avro Arrow. Although Velvet Glove was cancelled in 1955, the scientists at CARDE gained important experience with solid propellants. This experience was put to use when they wanted a rocket that could fly on short hops into the ionosphere for military communications experiments.

Their first rocket, called the Propulsion Test Vehicle, allowed for testing of various fuels and firing setups, but it was too heavy for research flights. The CARDE experts built a flying version that they called Black Brant after a fast Arctic goose. Two Black Brant I test vehicles flew successfully from the Churchill Rocket Range in northern Manitoba in the fall of 1959. Further test flights of early versions of the larger Black Brant II rocket built at CARDE were also successful. But the Canadian government had no interest in building rockets, and so CARDE contracted with a firm in Winnipeg, Manitoba, today known as Bristol Aerospace, to build more advanced versions of Black Brant for use by researchers as a sounding rocket to carry scientific instruments high in the atmosphere or into space, but not into orbit.

The man at Bristol Aerospace in charge of Black Brant was Albert Fia, a native of Lethbridge, Alberta, who joined the Canadian Army in World War II and took part in the D-Day invasion of France in 1944. After the war, Fia re-enlisted in the military and obtained an engineering degree at the Royal Military College of Science in England and Laval University in Québec City. In 1958, when Bristol wanted someone to direct their rocket efforts, they hired Fia, who by then had vast experience with weapons. "His military background required precision that's very important

when you get into the area of rockets and explosions
are happening," said Murray Auld, who was general
manager at Bristol when Fia worked there. Fia himself
was conscious of the fact that rocketry is an unforgiving
art—even the smallest mistake means an explosion or
a failure.

Fia and his special projects group at Bristol learned
that lesson the hard way. Fia's team built and tested the
6-metre-high Black Brant III rocket on test stands, but
when it came time to launch the rocket in 1962, the
Black Brants quickly lost their stability in flight. In spite
of changes made in later test vehicles, the stability problem continued to draw the rockets off course. While
Fia's team continued to deal with the problem, they
readied a two-stage Black Brant IV rocket for launch in
1964. An explosion rocked the vehicle when the stages
separated, but the rocket survived and continued to fly,
though not as high as planned. A similar problem
plagued the second Black Brant IV. Information radioed to the ground on the second flight gave the Bristol
team the cause of the explosions, and a redesigned
stage separation system led to success on the test flights
that followed. Bristol moved on to the Black Brant V to
succeed the troubled Black Brant III, but the new rocket
again met problems in the test stand. Changes to the
motor design and measures to improve the rocket's fins
from Black Brant III finally led to a string of successful
flights when the rocket was launched starting in 1964.
These successful launches put Bristol Aerospace—and
Canada—into the rocket business.

Launch of a Black Brant rocket

Since that time, the Black Brant family of rockets has racked up an unparalleled record—800 successes and only 29 failures in launches from 22 different launch sites around the world, including Churchill. "Its reliability was incredible—over 98 percent both in terms of its firing and its abilities to reach high altitudes for experiments. The Black Brant's staying power is all the evidence you need," said Paul Heide of Bristol Aerospace. The Black Brant was launched from Churchill to probe the aurora borealis and the ionosphere, and its trajectories allowed other experiments to fly to altitudes of up to 1500 kilometres or to undergo weightlessness for periods of up to 12 minutes. The sounding rocket, which was upgraded 12 times, has become an export success story for Canada and remains in use today.

As for Fia, he retired from Bristol in 1980, and the next year, NASA gave him its Public Service Award for his "dynamic leadership in the development of solid propellant motors and rockets," a rare award for a non-American. When Fia died in 2005, at the age of 89, he was saluted as Canada's "rocket man." In spite of Black Brant's sterling record, Canada's only research rocket remains a little-known part of Canada's efforts in space.

One reason is that the Black Brant could not fly into orbit, although Fia and his colleagues drew up plans to build such a rocket on more than one occasion. "We could have a Black Brant rocket capable of launching a small satellite within 10 years," Fia told a newspaper in 1967. "The only thing lacking is a customer." The Chapman report into Canadian space activities recommended that Canada build such a rocket, but the federal government never took up the idea, and as a result, every Canadian satellite has been launched on foreign-built rockets.

CHAPTER FOUR

Cannons of Controversy

IN ADDITION TO HELPING CREATE THE TECHNOLOGIES THAT LED TO Black Brant, the Velvet Glove program at the Canadian Armament Research and Development Establishment (CARDE) helped spawn the career of a controversial and daring rocket scientist, one whose efforts eventually ended in failure and death. In 1950, a young PhD student from the University of Toronto named Gerald Vincent Bull started work at CARDE on Velvet Glove. His job was to develop a proper shape for the missile's airframe that would allow it to fly correctly at supersonic speeds. Bull started his work while he completed his PhD thesis from the University of Toronto's newly established Institute of Aerophysics.

The institute, which later became known as the University of Toronto Institute for Aerospace Studies (UTIAS), was establishing itself as a major centre for research into high-speed aircraft. "We were well aware at that time, having solved the sound barrier problem, that we were riding the crest of a wave of development that clearly was leading man into space," said UTIAS' founding director, Gordon N. Patterson, who studied at the prestigious Institute for Advanced Studies at Princeton, New Jersey, the home of scientific luminaries such as Albert Einstein and John von Neumann. Patterson supervised Bull's studies, and UTIAS would go on to train many of the Canadians who helped build Canada's space program in the 1960s and beyond.

Bull researched using supersonic wind tunnels at the institute, building one in the university's engineering building. He then moved with the institute out to its long-term home in Downsview, north of Toronto, where he helped build a wind tunnel capable of supporting tests at seven times the speed of sound. When Bull got his PhD and moved full-time to CARDE in 1951, he was soon heading a research team looking at the aerodynamic capabilities of Velvet Glove. He needed to test the missile's aerodynamic characteristics above the speed of sound. There was no wind tunnel at CARDE, and the UTIAS wind tunnel wasn't suitable for the job.

The CARDE facility at Valcartier had long been used to test cannons and other artillery. The Québec site included a tunnel where artillery shells could be fired in simulated conditions at various altitudes. Bull had the idea of testing models of Velvet Glove by firing them from a cannon into the tunnel. Inside the cannon, each model was fitted inside a sabot (a band made of wood or metal that surrounds an object so it fits the barrel of the cannon).

Bull's idea to use the guns to test missile bodies was a great success, and he built further facilities at CARDE to test models at high speeds using hypervelocity guns. But more importantly for Bull, he became interested in large guns and the possibilities they offered. His research continued after Velvet Glove was cancelled in 1955. Bull not only developed guns that could fire shells faster and higher, but he also built shells containing instruments that could survive the shock of firing and send telemetry to ground stations as they flew. While Bull won promotions and pay increases for his groundbreaking work, he found that not all Canadian government authorities shared his enthusiasm. But U.S. military authorities, notably the head of the U.S. army's research and development, Lieutenant General Arthur Trudeau, began to follow Bull's activities and even visited Valcartier to see test firings of Bull's guns.

Finally, in 1961, Bull resigned from CARDE out of frustration with what he saw as a lack of support. He was unpopular with many of his colleagues after newspaper and magazine articles promoted him as Canada's "boy rocket scientist" and plugged his provocative idea of launching Canada's first satellite into space using a super cannon.

Bull's basic idea was nothing new. The French writer Jules Verne, in his books *From the Earth to the Moon*, published in 1865, and *Trip Around the Moon*, which followed two years, described a trip around the Moon by three men who rode inside a projectile that was launched by a gigantic cannon from Florida. The trip ended with a splashdown in the Pacific and a pickup by the U.S. navy. Verne's novels have been credited with firing the imaginations of the great rocket and space pioneers of the 20th century, and young Gerry Bull had read Verne. But Verne glossed over the great difficulties of building such a large gun—and the fact that no human or animal could survive being shot out of a cannon. While other space pioneers decided that the rocket was the way to get into space and to the Moon and the planets, Bull proposed using a cannon to send instruments into orbit. After a difficult childhood and apprenticeship in engineering, Bull had a dream, and he was setting out to pursue it.

At the time he left CARDE, Bull was only 33 and newly married. He was born on March 9, 1928 in North Bay, Ontario, the ninth child born to George L. Toussaint Bull, a lawyer of United Empire Loyalist, Anglican stock, and Gertrude LaBrosse, a French Catholic. When Gerry Bull was two years old, the family moved to Toronto after the Depression brought down his father's law practice. Early in 1931, Gertrude gave birth to a 10th child, but she died a few weeks later. The elder Bull couldn't cope with the loss of his wife and left the children in the care of his sister, who was stricken by cancer and died at Christmas 1932.

The older children were left to fend for themselves, and the younger children were left with relatives. When George Bull remarried in 1934, there was no place for his children in his new marriage. Young Gerry Bull's sister helped raise him for a time, but his mother's brother and his wife, who had no children but were financially secure thanks to a winning ticket in the Irish Sweepstakes, took up the task of raising their young nephew at their home in Kingston. Bull's aunt detested the boy's father and raised young Gerry apart from the rest of his family. She was also distant and demanding and sent Bull to boarding school at age 10. Gerry was quiet and withdrawn, and the tragedy of his life was magnified when a cousin he was close to was killed on active duty during World War II.

A gift of two model airplanes from his aunt and uncle fired Gerry's imagination, and he showed an aptitude for mathematics and science at his school in Kingston. When he graduated in 1944 at age 16, Bull wanted to go to medical school, but he would have to wait until he was 18. He could get into engineering at his age. Despite his youth, he impressed a professor with his knowledge of aeronautics and was accepted to study aeronautical engineering at the University of Toronto.

While Bull was working at CARDE and living in Québec City, he met and married Mimi Gilbert, the daughter of a local physician. Bull and his wife soon became known as good hosts to the researchers at Valcartier and their other friends. They also began their family of five boys and two girls. Bull was finally able to express the warmth to his friends and family that had been missing in his childhood.

A tall and solidly built man with a receding hairline but a handsome face, Bull was a hard-driving boss who expected top-flight work but allowed his employees a great deal of freedom. He worked hard to make and keep friends, and he enjoyed lightening the atmosphere at work with practical jokes. But he had little patience for those who stood in his way, particularly if they worked for government. A fellow engineer said, "He either made devoted followers or he made implacable enemies."

When Bull left CARDE in 1961, he fielded many offers, but he did not want to leave Québec because of his ties to his wife's family. In June, Bull joined McGill University in Montréal as a professor of engineering science in the department of mechanical engineering. At age 34, Bull was the youngest full professor ever appointed by the university. Bull and his family moved to Montréal and bought land south of the city on the border of Vermont, near Highwater, Québec, to use as a summer home and to continue ballistics testing. Bull and McGill formally founded the High Altitude Research Program (HARP) in an effort to use a cannon to launch a satellite into orbit.

Bull's work got support from the U.S. army, which was still smarting from a decision in the late 1950s that gave jurisdiction over space launches to the U.S. air force. Prior to that time, the army employed Wernher von Braun and his group of rocket experts, who built the V-2 rocket for Germany during World War II. That group of scientists had immigrated to the U.S. after the war. Von Braun's team built the Redstone and Jupiter missiles for the army and launched the first U.S. satellite, *Explorer 1*. But the U.S. government decision to take rockets and space from the army led to the departure of von Braun and his team for the civilian space agency, NASA. Bull's work with cannons gave the army a way back into space, but the army could only give him a small amount of help so as not to attract attention from higher authorities. In return, Bull helped the army by sharing his findings about how re-entry into Earth's atmosphere affected missile nose cones. He also conducted experiments into the use of pellets as a means of destroying incoming missile warheads. The pellets would be fired at incoming missiles, and the high-speed collision would destroy the warhead before it could reach its target. The idea is still being pursued as a part of U.S. missile defence research.

Bull and McGill sought funding from the Canadian government for HARP, but Bull's many critics at the Defence Research Board blocked funding from that source and frustrated efforts to get money from other governmental departments. One official claimed that HARP "consisted merely of large and expensive gadgets." Many scientists wondered whether instruments onboard a satellite could survive the shock of being fired from a cannon and questioned whether the small satellites proposed by Bull would be of much use. Nevertheless, Bull had strong backing from McGill and its dean of engineering, Donald L. Mordell. In 1962, Bull and Mordell held a press conference to publicize their plans. Beside them on the stage were three small vehicles designed to be launched from a cannon. The rockets were called Martlets, after the mythical heraldic bird that adorns the crest of McGill University. Bull suggested that he wanted to put one in orbit by 1967, when Canada would celebrate its centennial year. In interviews with reporters at the time, Bull said he didn't expect Canada to carry out massive research programs: "But I do see us operating in highly imaginative fringe areas, coming out with novel ideas and revolutionary-type thoughts." Bull also had some shots for his antagonists in government, bemoaning the fact that many Canadian scientists were compelled to work in the U.S. "Canadians won't gamble on research unless it's aimed at earning a dollar," he added.

Bull stopped his research on campus when a malfunctioning gas gun damaged a building. He moved the work out to Highwater. Originally, Bull envisioned testing cannons in northern Québec, but Mordell suggested Barbados, a Caribbean nation where McGill had research facilities in other fields. The southerly location was well suited for HARP, because the island's proximity to the equator gives vehicles a speed advantage on their way to orbit. Bull's friends in the U.S. army were able to procure a surplus naval cannon for his research, and the government of the newly independent Barbados gave its support because of the economic activity HARP could generate.

Bull built the Martlet 1 glide vehicle in 1962, and in Barbados the following January, he began firing the cannon he named Betsy. The final shot in that series, on February 1, saw a Martlet 1 fly 27 kilometres high and send back telemetry from a small transmitter that survived the launch. Within a few months, a Martlet 2 glide vehicle had reached 92 kilometres high. A series of Martlet 2s released gas or radar chaff into the upper atmosphere that scientists could follow from the ground to learn about currents at those high altitudes. Other Martlets sent back readings from small instruments that provided scientists with a wealth of data on the ionosphere. Military Martlet 2 shots released pellets for missile defence tests.

These early successes brought HARP substantial financial support from the U.S. army, and this in turn brought the Canadian government on board. Both governments agreed to fund HARP for three years, starting in 1964. McGill set up the Space Research Institute to run HARP, with Bull as director. With his new budget of nearly $3 million a year, a huge sum at the time, Bull was able to extend the barrel of his cannon and hire engineers to research solutions to the project's problems before he could build a Martlet rocket that could carry a Canadian satellite into orbit. He planned to use Martlet 3 rockets that would begin to fire after being shot from the cannon. With his projected Martlet 4 rocket, Bull planned to put a small instrument-carrying satellite into orbit. But the new rockets required the development of a guidance system that could operate following a cannon launch.

Among the new staff hired at HARP was Bruce Aikenhead, an engineer from the University of Western Ontario, who had worked on the Avro Arrow before its cancellation in 1959. He then went to train astronauts for NASA before returning to Canada in 1962.

"Gerry Bull was an absolutely enthusiastic super salesman," Aikenhead remembered. "I talked to Gerry and he was really enthused about what he could do.

He was interested in what I could do and it looked like a whole lot of fun. So I made the switch and I joined the HARP program as a systems engineer. I was put in charge of the Martlet 4 projectile."

Aikenhead explained that other groups were developing techniques to launch the Martlet 3 and Martlet 4 rockets from cannons so that the rocket fuel would not be damaged when the cannon fired it with a force of roughly 10,000 times that of gravity.

"He had another group making a hollow steel projectile which contained another projectile. It was, in effect, the first stage, and then the nose would come off and the rocket inside—the second stage— would ignite. In the nose of the second rocket was the payload. This was the Martlet 4 concept. All our studies showed we could attain orbital velocity. The analysis showed you could put a 20-pound payload in orbit for a fraction of the normal launch cost," Aikenhead said.

"We took the Martlet 4 to Barbados and it was fitted with a dummy rocket with inert propellant. The analyses showed we were going down the right track. It was a very nice swords-into-plowshares kind of thing. He and his people were taking weapons and converting them into research tools. But in the process, they were learning how to make better weapons. That's why we had such great cooperation from the U.S. military. We were able to buy expired propellant and other things at bargain prices."

Bull and his team did not have to wait long for signs of trouble after HARP was approved for funding. Payments from the Canadian government were slow in coming, delaying work on the Martlet 4 and its satellite. As the funding entered its third year in 1966, the Canadian government announced that its support for HARP would end on June 30, 1967, the eve of Canada's 100th birthday. The Canadian decision meant that U.S. funding would also dry up. Bull tried to restore funding by talking to reporters about the possibility of putting a satellite in orbit to cap Canada's centennial celebrations. But his efforts failed.

"We had simply started the hardware building too late to beat the deadline of the group known generally as 'The Torchers.' This particular organization acts as an arm of those dedicated to destroying technical projects," a bitter Bull wrote later.

"It looked to be a promising thing, and if it had not been cancelled, I'm sure we would have made orbit," Aikenhead recalled. "But Gerry had a tendency to jump from one project to another. Research and development funds were drying up on both sides of the border, and it was the Avro Arrow all over again."

As HARP was going through its final months, John Chapman and his colleagues completed their report on Canadian space efforts, but while their report contained an extensive description of HARP, the discussion of the program's future was carefully phrased. The Chapman report called for a small Canadian-built launch vehicle to send small research satellites into

orbit and suggested that the government study proposals involving HARP or a cluster of Black Brant
rockets to do the job. But the government never followed up on the suggestion. To this day, no satellite
has ever been put in orbit atop a Canadian rocket, and
no satellite has ever been put into orbit by a cannon.

As for Bull, he left McGill and converted the institute he ran into the Space Research Corporation (SRC).
Thanks to the early backing of people like the Bronfman
family of Montréal and some stock deals that paid off,
Bull was able to gain control of the properties he
inherited from HARP, including the then-10,000 acres
that straddled the border near Highwater, and continued use of the Barbados facilities. But in general, Bull
would prove ill suited to be a businessman after
having spent his life in government and academia as
a research scientist.

For one thing, Bull was entering a difficult business.
He was competing against huge armament firms that
had close links to the governments that were their
major customers. Many governments saw having
domestic arms suppliers as a matter of national security. They were reluctant to seek outside help, and then
only for as long as necessary. And in the late 1960s and
the 1970s, a growing anti-war movement pushed the
arms industry deep into the shadows. Also, the successful conquest of the Moon by Apollo astronauts
marked the beginning of a sharp reduction in money
available for aeronautical and space research.

Although the Space Research Corporation got off to a good start with some aeronautical research contracts, in the 1970s, Bull and his firm found that they had to survive on another source of business.

"He was able to stay afloat because countries employed him to solve their minor artillery troubles," wrote Bull biographer William Lowther. "Iran wanted to know why their new shells were tumbling rather than spinning towards the end of flight. Taiwan was puzzled by erratic performance from its 155-millimetre guns. Bull became 'Mr. Fix-it' for the world's artillery." Lowther estimated that in 10 years, Bull did work for no fewer than 30 countries.

Bull began hiring people to help him with work at Highwater, and there were so many visitors and workers there that a hotel was set up to deal with them all. When Bull was not travelling the globe to drum up business, he was entertaining visiting delegations at his family mansion on the Highwater site. SRC paid to have customs inspectors on both sides of the compound, but Bull was not afraid to take advantage of the differing laws and policies between Canada and the United States to advance his business. He shipped goods and services out the north or the south exit of his personal free-trade zone at Highwater as politics or the law demanded. Bull continued to do business with the U.S. military, but the fact he was not an American was becoming a problem. So in 1973, a bill sponsored by Arizona senator and former presidential candidate Barry Goldwater gave Bull his U.S. citizenship.

Bull also established a branch operation in the Belgian capital, Brussels, a well-known centre for arms trading. In the 1970s, the Cold War between the U.S. and Soviet Union was still in full swing. The two sides did much of their fighting in the form of proxy wars involving allies in the Middle East or in underdeveloped countries. Bull supplied expertise for Israel during this time. Through American and Israeli connections, Bull was drawn into one of the era's larger proxy wars, in the impoverished central African nation of Angola. The country had just emerged from a lengthy period of colonial domination by Portugal, and two major groups were contending for power. The Soviet Union and Cuba backed one, while South Africa supported another—with assistance from the U.S.

The white supremacist apartheid regime then in power in South Africa was under an arms embargo backed by the United Nations, but the country had a vigorous domestic arms program that included efforts to develop its own atomic bomb. South Africa also wanted to improve its artillery for use in Angola and elsewhere, so they hired Bull to work on improved shell casings and better gun barrels. The shell casings from SRC soon caught the attention of governments and investigative journalists.

A change of government caused Bull to close his operation in Barbados and shift it to nearby Antigua, but news of Bull's work for South Africa caused troubles there, too. Canada's Prime Minister Pierre Elliott Trudeau, who strongly opposed apartheid, ordered an investigation of Bull's activities. The U.S.

government also became involved when President Jimmy Carter, who was concerned with human rights, moved into the White House in 1977. Although Bull believed that his shipments to South Africa followed U.S. law, American prosecutors began an investigation. After a lengthy proceeding in front of a grand jury, U.S. prosecutors charged Bull and an associate on March 25, 1980, with breaking the U.S. arms embargo to South Africa. On the advice of their lawyers, the two men pleaded guilty. When they were sentenced to a year in jail with half the sentence suspended, Bull was shocked because he had not believed he would be sent to prison.

As a result of the convictions, Space Research Corporation was shut down, and its assets were sold to deal with creditors. The company paid fines to cover similar Canadian charges of breaking the arms embargo. Bull and his wife were able to hang on to their two homes and an inheritance Bull had recently received from his late aunt. But as he was faced with the loss of his life's work and the prospect of jail, Bull's mental condition deteriorated, and he began to drink heavily. He was hospitalized to deal with his psychiatric problems, which were rooted in his difficult childhood. On October 1, 1980, Bull entered the minimum security Allenwood Prison Camp in Pennsylvania and served 4 months and 17 days, with remaining time taken off for good behaviour. Although the psychiatric treatment and the relatively relaxed atmosphere of the jail did him good, Bull emerged angry with the American and Canadian governments, and he was still under a cloud of negative publicity about his activities.

After a tropical vacation with his wife, Mimi, Bull decided to return to work and base himself in Europe. He reconstituted the Space Research Corporation and began to pursue a new contact that led him to do business with the People's Republic of China. Many people who knew Bull were puzzled by this turn of events because Bull was strongly anti-Communist. But Bull needed to get his business going again, and he was still bitter at what he saw as an unfair jail sentence for helping out an unpopular Cold War ally of America. In the early and mid-1980s, Bull rebuilt his business and helped the Chinese develop their own sources of artillery. Although Bull lived apart from his wife, she often came to his new home in Europe to visit, and he also made regular visits to Montréal. And Bull brought his sons Michel and Stephen into his business. The well-read Bull also found time to write a book about HARP and the Paris Gun, the German cannon that advanced artillery during World War I. Thanks to the ongoing political and military turbulence in the Middle East, Bull would get another chance of pursuing his dream of a supergun.

One of the bloodiest wars of the time involved Iran and Iraq. In 1979, an Islamic revolution toppled the Shah of Iran, and Iraqi dictator Saddam Hussein decided to exploit the turmoil by attacking Iran to grab territory that he coveted near the Persian Gulf. Hussein knew that Iran's officer corps had been weakened by the revolution. He was also aware that Iran was deeply unpopular in the United States because it had held the staff of the U.S. embassy there as hostages between 1979 and 1981 and had urged other acts of defiance against the U.S. and other Western countries. But Hussein had not bargained on the revolutionary fervour that gripped Iran. Waves of Iranian soldiers gave up their lives as they faced Iraqi arms in what they saw as a holy war. When Iraq suffered serious setbacks in the war, it obtained assistance from the administration of U.S. President Ronald Reagan, who was anxious to keep Iran isolated and weak. Hussein also got help from the Soviet Union, which supplied him with primitive short-range Scud missiles. Iraqi engineers upgraded the missiles, and with the other help Hussein received, Iraq was able to drive back the Iranians and end the war in 1988.

In 1987, just as Bull's business with China was winding down, Hussein began aggressively seeking out technology that could give his military an edge. The effort paid off with the Scuds that hit Iranian cities and helped bring the Iran–Iraq war to an end. Iraqi government officials and businessmen soon got in touch with Bull and brought him to Iraq. The Iraqis asked Bull for help, and they responded positively when Bull spoke of his dream of a supergun that could put a satellite into orbit.

Bull never gave up the dream he had been working on during the days of HARP. In the 1970s, he hoped to get the U.S. military interested in building a huge cannon, and his dreams were renewed in 1983 when Reagan announced a massive effort to build an anti-missile

shield for America under the planned Strategic
Defence Initiative, which was popularly known as
Star Wars. As the Iraqis bargained with Bull to help
them with their artillery and missiles, they also agreed
to use their oil wealth to help Bull build his supergun
in Iraq. It would be called Project Babylon.

The gun would be built in the side of a mountain
and could be aimed in only one direction. It would be
easy to see from the air, so it would be of little use as
a weapon. With his dream suddenly back within
reach, Bull made sure that his contacts in the American, British and Israeli militaries knew about Project
Babylon and its limited utility as a weapon. None of
his contacts objected to Project Babylon.

But as happened when he worked with South
Africa in the 1970s, Bull became caught up in changing political currents. The end of the Iran–Iraq war
caused the U.S. and other Western countries to look
more critically at Hussein's bloody suppression of
human rights in Iraq. And Hussein strongly hinted
that he wanted to take on Israel and establish himself
as the most popular figure in the Arab world. Israel
looked on Hussein's efforts to upgrade his military
with a wary eye. In 1981, Israel bombed Iraqi nuclear
reactors because of concerns that Hussein wanted
nuclear weapons, and the Israelis faced down the
condemnation that followed.

By 1989, Bull established his home base in an apartment in the suburbs of Brussels. That spring, as he
worked on his projects with Iraq, Bull was disturbed

by signs that intruders were visiting his apartment while he was away. He left large amounts of cash there, but it was never touched. The intrusions on his apartment ended in the summer, but business associates began receiving word that Bull's life was in danger.

In October, Bull met alone with Israeli officials. Although he never told anyone what was discussed, many believe he was warned about Israel's concerns about him helping Iraq with artillery and missiles. Two months later, Iraq attempted to launch a satellite using a rocket assembled from Scuds. Bull had helped with the design. Everyone knew that such a rocket could also be used to deliver weapons beyond the limited range of the Scud missile. Saddam would likely use such a weapon against Israel. The international community's concerns were later proven when he fired Scuds at Israel during the 1991 Gulf War.

Early in 1990, Bull visited Canada and the U.S. with his wife. He also visited China to do some consulting work. People who saw him at the time say that he appeared exhausted and troubled. On March 22, 1990, Bull's personal assistant dropped him off in front of his Brussels apartment building. When Bull emerged from an elevator on the sixth floor and walked to the door of his apartment, an assailant stepped from the shadows and killed Bull with five bullets from a revolver fitted with a silencer. Although the killer has never been found, it is widely believed that Bull's death was ordered by the Israeli Mossad secret service after Bull failed to respond to the organization's warnings about his work for Iraq. Some have suggested that the U.S.

Central Intelligence Agency or other covert groups may have been involved as well.

In the weeks following Bull's death, a shipment of precision-engineered tubes destined for the Project Babylon supergun was detained in England, but an inquiry was halted because the British government had approved the export at a time when Hussein was seen more favourably by the British government. Iraq's appetite for technology was already in the news because of another seizure of devices destined for Iraq that were normally used in nuclear weapons. The following August, Hussein invaded Kuwait, an act that touched off the Gulf War of 1991. Hussein remained in office until American and British troops invaded Iraq again in 2003.

As for Bull, he was buried near Montréal. His funeral was attended by more than 600 people. Although he has gone down in history as an arms trader, he gave up his life while pursuing his dream of a cannon that could launch satellites into space. His drive to reach his goal in the face of harsh political realities and people who didn't share his enthusiasm makes Gerald Bull a unique figure in the history of the quest to go into space.

CHAPTER FIVE

The Avro Arrow

ON FEBRUARY 20, 1959, A FRIDAY MORNING IN THE MIDDLE of winter, thousands of people inside the Avro Canada plant alongside what today is Toronto's Pearson International Airport were hard at work building Canada's most advanced jet aircraft, the CF-105 Avro Arrow. The first Arrow, a gleaming white aircraft whose name suggested its shape, emerged from the plant nearly a year and a half earlier. Since then, it and five other Arrows had taken to the air, flying at speeds near twice the speed of sound. Just the day before, Avro test pilot Wladyslaw "Spud" Potocki had flew two Arrows in short flights. At the head of the assembly line that Friday was the sixth Arrow, the first that was equipped with Canadian-built Iroquois engines that would allow it to fly even faster and farther than the first five Arrows.

Although the Arrow was a source of pride for the people who built it and for many other Canadians, it was also the subject of political controversy. The federal government under Prime Minister John Diefenbaker was concerned about the growing costs of the Arrow. The previous September, Diefenbaker had announced that the Arrow program was under review. His government tried to sell the Arrow to the U.S. air force and other foreign buyers, but it had no takers. Indeed, the U.S. government offered aircraft of its own that cost less than the Arrow, and a missile, the Bomarc, that was designed to shoot down hostile aircraft. On top of everything else, there were questions about whether the Arrow could effectively defend Canada. At the time, Canada, the United States and their allies were locked in an uneasy stalemate with the Soviet Union—the Cold War—and both sides possessed nuclear weapons. The Arrow was designed to intercept aircraft carrying nuclear weapons from Soviet Russia over the North Pole and Canada to their targets in the U.S. But in 1957, the Soviet Union developed an intercontinental ballistic missile that could deliver nuclear bombs anywhere on Earth. No aircraft could defend against such a missile. Military leaders did not know at the time whether there would still be a role for aircraft like the Arrow.

At 11:15 that morning, loudspeakers around the Avro plant interrupted work with the news that Prime Minister Diefenbaker had just announced in Ottawa that the Arrow program was cancelled, but that work would continue until further details could be obtained. That afternoon, an upset Crawford Gordon, the president of Avro Canada, came to the microphone to tell his 14,000 employees that they were laid off immediately. All of the staff members except the security guards were told to put down their tools and leave the plant immediately. Except for the few workers who came back to shut down the Arrow assembly line and those engaged in a futile effort to find something to make instead of the Arrow, Black Friday marked the end of the Arrow program—and Avro. Within two years of Black Friday, Avro Canada went from one of the largest corporations in the country to oblivion.

Among the people closely identified with the Arrow was a lanky, bespectacled, 43-year-old engineer named James Arthur Chamberlin, who was then Avro's chief of technical design. For almost his whole life, he had a passion for aircraft, and he spent most of his adult life at or near the summit of Canada's aircraft industry. Chamberlin was born in Kamloops, BC, on May 23, 1915. His father, Arthur Chamberlin, joined the Canadian army to fight in World War I and lost his life at Vimy Ridge in 1917. The young Chamberlin spent his first few years in BC until his mother remarried, and the family moved to Toronto.

"Almost as soon as he could walk, Jim was building airplanes," his mother recalled. The young Chamberlin was more interested in designing aircraft than in flying them. As a teenager, he won a model aircraft competition sponsored by a Toronto newspaper. After going to high school in Toronto, he studied mechanical engineering at the University of Toronto and then went on to study aeronautical engineering in England at the Imperial College of Science and Technology, where he earned a master's degree just as war was about to break out in 1939. During World War II, Chamberlin put his aircraft design talents to work on the Canadian version of the Avro Anson and the Noorduyn Norseman. During the war, Chamberlin married Ella, and they had a daughter and a son.

The Canadian government established Victory Aircraft Ltd. during the war to build Lancaster bombers to add to the Lancasters that were already being built in Britain. Victory Aircraft set up in a former railcar plant in the Toronto suburb of Malton. After the war, the government handed over the plant to the British firm Hawker-Siddeley Aircraft. Hawker-Siddeley's legendary head, Sir Roy Dobson, set up a Canadian company, Avro Canada, which began assembling Canadian capital and talent to build new aircraft. One of Avro's first hires in 1945 was Jim Chamberlin.

Avro Canada set to work on two new aircraft. One was a jet interceptor, the CF-100 Canuck, that became a mainstay of the Royal Canadian Air Force in the 1950s and 1960s. The speeds and technologies of jet aircraft were new to everyone at the time. Avro's engineering team, under James Floyd, dealt with many difficult problems and learned hard lessons while building the CF-100. As chief aerodynamicist, Chamberlin faced many daunting problems with the aircraft.

At the same time, Avro Canada designed and built North America's first jet transport, the C-102 Avro Jetliner. When the Jetliner took off from the Malton airport for the first time on August 10, 1949, it was far ahead of anything in the U.S., but it was two weeks behind the British de Havilland Comet in the race to be the world's first jet transport. When the Jetliner flew to New York a few months later, it received heavy press coverage. American multimillionaire Howard Hughes, who owned, among other things, an airline, took the controls of the Jetliner and wanted to buy it.

Within a few days of the Jetliner's first flight, the Soviet Union exploded its first atomic bomb. In 1950, the Korean War broke out, with Western nations supporting South Korea, and the Soviet Union and Communist China backing North Korea. The CF-100 made its first flight early that year, but the aircraft was in trouble, and it needed a major redesign. C.D. Howe, the powerful Liberal cabinet minister who ran Canada's defence production and transportation departments during and after World War II, ordered Avro to concentrate on defence duties. Avro duly fixed the CF-100, but the decision doomed the Jetliner project. Only one Jetliner had been built, but 692 CF-100s rolled out of the Avro plant during the 1950s, including 53 that were sold to the Belgian Air Force. To this day, the CF-100 remains the only Canadian-designed military jet aircraft to be put into mass production.

By the time the CF-100 program got on track, the RCAF decided it would need something that could outfly the subsonic aircraft. Indeed, it wanted something that could fly twice the speed of sound, operate in Canada's harsh Arctic environment and maneuver at high altitudes without losing speed. So Floyd, Chamberlin and others on Avro's growing engineering team started designing the aircraft that eventually became the Avro Arrow.

The Avro team incorporated the latest advances into the delta-winged Arrow, including its production methods and its aerodynamic shape. Each Arrow carried a modular weapons pack that could be quickly changed out, an air-conditioning system to protect the aircraft electronics and pilot from the high temperatures of supersonic flight and, a first, a fly-by-wire system that allowed the pilot to control the aircraft with the help of an automated system. Both the Arrow and its Iroquois engine used the latest metals, such as titanium and exotic alloys, and they required the latest machinery to produce their parts. The Avro plant became the home of one the first examples of a new device, an IBM mainframe computer, which was needed to do many of the calculations required to produce the design. The computer also helped operate a simulator used to train pilots and predict how the aircraft would behave in flight.

To test their aircraft design, Chamberlin and other engineers became frequent visitors to the United States. There they flew American rockets topped with models of the Arrow after flying similar rockets in Canada, and they examined the Arrow's flying characteristics in gigantic American wind tunnels. The Avro designers consulted with top experts at the U.S. National Advisory Committee for Aeronautics, which developed and tested the latest advances in aircraft science and, in 1958, formed the core of America's new space agency, NASA.

The first CF-105 Avro Arrow during a 1958 test flight

Throughout the Avro Arrow's development, C.D. Howe kept a critical eye on the new aircraft and expressed concern about its growing costs. Still, he supported the project. But voters retired him and the rest of the Liberal government in the 1957 federal election that swept Diefenbaker and his Progressive Conservatives to power in Ottawa.

Three months after the new government took office, Avro Canada scheduled an elaborate ceremony to unveil the Avro Arrow. A crowd of 15,000 people, including Avro workers and their families, joined dignitaries from various governments as the first Arrow rolled out of the plant on the afternoon of October 4, 1957. At parties after the ceremony, radio reports broke in with the news that the Soviet Union had launched the Earth's first artificial satellite, *Sputnik 1*. The news shocked most people outside the Soviet Bloc, and so did the news a month later that the Russians had launched an even bigger satellite with a dog on board. The launch of the Sputniks served notice to the West that the Soviet Union possessed rockets that could deliver nuclear bombs anywhere on Earth in minutes. The news got worse in December, when the first American attempt to launch a satellite ended in a ball of flames on a Florida launch pad.

Although Chamberlin, Floyd and the others at Avro knew that the space age had dawned, they still needed to prove their new aircraft. On March 23, 1958, legendary Avro test pilot Janusz Zurakowski took the Arrow on a successful first flight. Most of the flights that followed were equally successful, although some flights saw trouble from parts such as the aircraft's advanced landing gear. Also, the first five Arrows flew with American engines that were not as powerful as the Iroquois engine that was nearly ready to be joined to the sixth Arrow on Black Friday.

During that year of test flying, Chamberlin and his team used the test pilots' findings about the CF-105 to improve the aircraft coming off the assembly line, and they began designing even more advanced versions of the Arrow, including versions that could fly to the edge of space. Other engineers at Avro worked on another cutting-edge aircraft, a flying saucer known as the Avrocar. The Avrocar was built for the U.S.

military, but it contained many design defects that limited its flying abilities. The Avrocar continued to fly after the Arrow's cancellation, but the project died two years later with Avro Canada.

The Avrocar employed only a few people. For most people at Avro, Black Friday meant the end of work on the Arrow and the beginning of job searches. Within days, American aircraft firms quickly snapped up many of the top engineers from Avro. Others found engineering jobs in Canada, and some were forced to start from scratch in new fields. But a small group of Avro engineers was soon going to join a program few of them knew about, facing challenges they had never contemplated.

CHAPTER SIX

Jim Chamberlin and Gemini

LIKE MANY PEOPLE AT AVRO, JIM CHAMBERLIN AND AVRO'S CHIEF engineer Bob Lindley knew many people at NASA from their use of U.S. wind tunnels to test models of the Arrow. By the time the Arrow project was cancelled, NASA was setting up a program to send the first astronauts into space. Chamberlin and Lindley called their contacts at NASA to see if the space agency would be interested in a loan of engineers from Avro while the Canadian firm looked for new work. Both Lindley and Chamberlin flew to Washington to talk to NASA officials. The idea of a loan was quashed, but the agency made it known that it was short of engineers and wanted to hire some of the top people from Avro. Lindley and Chamberlin summoned Avro's engineers to the plant cafeteria and told them that NASA was looking for people like them.

Three weeks after Black Friday, a plane carrying top NASA officials, including Robert Gilruth, the director of Project Mercury, landed at the airfield next to the Avro plant. The NASA officials got a good look at the five completed Arrows, which a few weeks later were broken up for scrap. One of them remembered thinking that the Arrow was "superior to anything the United States had at the time." The next day, Gilruth and his colleagues interviewed 100 Avro engineers and offered jobs to 25 of them, including Chamberlin. Lindley didn't apply, instead opting to remain with Avro until it formally folded. In the months that followed, NASA hired another six Avro engineers. Others, including Lindley, later became involved in the U.S. space program when U.S. aerospace firms hired them. Lindley finally joined NASA in 1969.

In April 1959, Chamberlin and his 24 colleagues from Avro reported to work at NASA's Langley Research Center in Virginia, where just months earlier they had tested Arrow models in the facility's gigantic wind tunnels. Langley was the home of the Mercury program until a new space centre was set up in Houston, Texas, in 1962. Mercury was such a new program at the time that many of the Avro engineers reported for work the same day as seven special recruits who would fly the new spacecraft as astronauts. One of the Avro engineers, Bruce Aikenhead, went to work training the astronauts using a variety of devices designed to simulate the conditions in space.

As for Chamberlin, he was given one of the top jobs in the Mercury program: head of the engineering division and responsibility for the office that ensured that Mercury's manufacturers built each spacecraft and component according to NASA's specifications. Maxime Faget, the legendary engineer who designed spacecraft from Mercury to the space shuttle, said Chamberlin was put "in charge of the day-to-day management of the Mercury program." Building a vehicle to carry astronauts into space was something completely new and different from anything people had done before; Jim Chamberlin had been given one of the most challenging jobs in engineering.

Not only was the Mercury spacecraft new, but so were its rockets. The Redstone rocket that carried Mercury prototypes on suborbital hops into space had to be modified so it could safely carry astronauts. Many rockets shake violently during powered flight, and the flight would have to be made smoother if astronauts were to climb on board. The Atlas rocket that was to carry Mercury into orbit had a bigger problem—it had a habit of exploding in flight. But Mercury and its rockets gradually overcame a series of problems, and in April 1961, final preparations were under way for the first flight with an astronaut on board.

Then, on the morning of April 12, word came from Moscow that the Soviet Union had launched an air force officer named Yuri Gagarin into orbit. Gagarin made it home safely after a single circuit of the Earth. The U.S. was upset at being beaten again by Soviet rockets. When Mercury astronaut Alan Shepard made a suborbital jaunt into space three weeks later, he received a hero's welcome by the Americans, who were proud to be back in the race.

President John F. Kennedy watched these events and decided that America needed to become the world's top space power. Just 20 days after Shepard's 15-minute flight, Kennedy set a national goal for the

United States: "I believe that this nation should commit itself to achieving the goal, before this decade is out, of landing a man on the Moon and returning him safely to the Earth."

Even though the U.S. Congress quickly increased NASA's budget, the space agency had a daunting task. It had flown only one astronaut 170 kilometres high, and now it has eight-and-a-half years to get astronauts to the surface of the Moon, 400,000 kilometres away, and then get them back home safely. Not only did NASA have to build new rockets and spacecraft for the task, but it also had to train astronauts and flight controllers to make the flight successfully.

The program that would send astronauts to the Moon was called Apollo. Gilruth and others at NASA were worried that the leap to Apollo from Mercury, which was so small and limited that it was often called a space capsule rather than a spacecraft, was too big. Even before Kennedy set his lunar goal, Chamberlin had the job of making changes to Mercury so that it could fly longer than just a few orbits. Chamberlin decided that Mercury wasn't up to the task. A whole new spacecraft was needed, and he set to work designing one that could carry two astronauts. The new spacecraft was named Gemini, and it would be able to change orbits and allow astronauts to step outside and do spacewalks. Gemini was also designed to join up with a rocket named Agena that could boost it into higher orbits.

While Chamberlin was working on Gemini, he was also drawn into the arguments raging inside NASA about how to get to the Moon. There were three main ideas about how to get there. The most popular idea at first was the direct method: put astronauts in a rocket that would leave the Earth, carry them directly to the Moon's surface, and then bring them back to Earth. The second idea involved using two rockets to assemble a lunar spacecraft in Earth orbit (Earth orbit rendezvous). The assembled vehicle would then head directly to the Moon and bring the astronauts home afterward. The third idea, at first the least popular, was called lunar orbit rendezvous. It involved launching a mother ship and a small lunar landing craft from Earth into orbit around the Moon. The astronauts would transfer to the landing craft, land on the Moon and then return to the mother ship, which would then bring the astronauts home.

While the direct method made the most sense at first glance, it required a rocket that was much bigger than what would be needed for the other methods because the craft would need to carry enough fuel to get the whole spacecraft to the Moon and then launch back off the Moon's surface, including all the gear for the return trip to Earth. Designing a single spacecraft to do everything would be difficult, something like designing an amphibious car for a fishing trip.

A benefit of lunar orbit rendezvous was that the lunar landing tasks could be separated from other parts of the flight, such as Earth launch and re-entry. Lunar orbit rendezvous lowered the weight of the spacecraft

James A. Chamberlin, manager of NASA's Gemini program, sits behind a model of Gemini in his office in Houston in 1962.

and required a smaller booster rocket because only the lunar landing craft would need to descend to the lunar surface and leave again. The equipment to return home would stay in the mother ship in lunar orbit. Instead of the amphibious car, this concept was more like using a car, boat and trailer for a fishing trip. But the idea of astronauts having to find and join up with another spacecraft when they were 400,000 kilometres away from home terrified many people at NASA.

A NASA engineer named John Houbolt was convinced that lunar orbit rendezvous was the best way
to go to the Moon, but few people would listen to
him. Maxime Faget strongly criticized him at a meeting. And worse, Houbolt's job was not directly related
to the Moon flights. In a series of meetings, Houbolt
endured the criticism, until one day, when one of the
people involved with the Apollo decision, Jim
Chamberlin, asked for his figures. Chamberlin was
won over, and he decided to announce his conversion
in a dramatic way.

Chamberlin set to work on a bold idea. Why not
launch Gemini on a Saturn rocket along with a "bug"
that could carry a single astronaut to the lunar surface?
The bug was a tiny rocket vehicle with an open cockpit
that would fly like a small helicopter. He took the idea
to his boss, Robert Gilruth. Gilruth told Chamberlin to
halt his work on Gemini lunar flights and concentrate
on getting Gemini ready for its more limited goals in
Earth orbit. But he admitted that Chamberlin had
made a convincing argument for lunar orbit rendezvous for Apollo. Soon Gilruth and others, even Faget,
joined the ranks of NASA engineers in favour of lunar
orbit rendezvous. There were still months of debate
ahead, but finally, in July 1962, NASA announced that
it would use lunar orbit rendezvous to get Apollo to the
Moon and back. NASA set to work building a mother
ship, the command and service module, that would
carry three astronauts to lunar orbit and back to Earth,
and a smaller craft, the lunar module, that would carry
two of them to the surface of the Moon and back to

lunar orbit. Jim Chamberlin played a key role in one of the most important—and most expensive—decisions in NASA's history.

While NASA was debating how to get to the Moon, Chamberlin was busy designing his new spacecraft. Even though his idea of sending Gemini to the Moon didn't fly, he convinced NASA that the new spacecraft could do a better job of testing new flight methods than the improved Mercury capsule NASA planned to use. Late in 1961, Gilruth announced that astronauts would fly aboard Gemini spacecraft after the Mercury program ended and before Apollo took to the skies. He named Chamberlin as Gemini's project manager.

At both Avro and NASA, Chamberlin was known as a no-nonsense engineer who wasn't afraid to argue over technical points. His former boss at Avro, Jim Floyd, occasionally clashed with Chamberlin but remembers him as "absolutely a genius." Another colleague remembered that Chamberlin "was very smart but didn't have the appearance of being smart." In the workaholic world of NASA in the 1960s, Chamberlin stood out for his dedication to his job. He was not known to make small talk or chat about his family at work, but those who knew him well learned he enjoyed a good joke and liked sports such as sailing and swimming. And when he was at home, he did not let his workday concerns interfere with his family responsibilities.

In his new job, Chamberlin was immediately faced with a series of technical problems and budget issues.

Gemini was designed to be put into orbit atop the air force's new Titan II booster. Like other rockets, the Titan II needed design changes because its ride was too rough for astronauts. The problem, described as pogo because it involves up-and-down motions that results from uneven fuel flow in the rocket and feels like a pogo stick ride, proved difficult to eliminate. Gemini was originally designed to land in the desert under an inflatable wing called a paraglider. But the paraglider's development problems were bigger than Gemini's schedule permitted, so it was dropped in favour of a parachute deployment followed by a splash-down in the ocean and a pickup by the navy, the same method that was used in Mercury and Apollo.

Other advances also took time and cost money. Gemini included many innovations from Mercury, including an onboard computer; a new power source, fuel cells, to replace Mercury's batteries; new thrust-ers; and ejection seats to carry astronauts away in case the Titan exploded on the launch pad. The layout of Gemini was changed to allow easy changeout of trou-bled equipment on the launch pad. As well, the Agena rocket, in effect a giant fuel tank, had to be tested before Gemini could join up with it in orbit.

These new technologies caused Gemini's budget to rise even as the U.S. Congress was cutting back NASA's budget requests. The air force was impressed with Gemini, and NASA officials had to hold back military efforts to take over the program. By the spring of 1963, Gemini was over budget and behind schedule, and NASA decided to replace Chamberlin as project manager, although he stayed on with the agency. "If we had waited to get accurate cost estimates before proceeding, we could not possibly have beaten Apollo," Chamberlin said later. "The rise in Gemini costs was nothing like those in Mercury or Apollo." Others said the change made sense. Chamberlin's strength was in design, and his replacement was an operations expert who could direct the Gemini flights.

A year later, the first Gemini test rose off the launch pad without astronauts on board. Then, on March 23, 1965, America's first crew of astronauts, Virgil "Gus" Grissom and John Young, climbed aboard *Gemini 3* and put the spacecraft through its paces. Another nine flights followed during the next 20 months. The highlights included Ed White's spacewalk on *Gemini 4* and *Gemini 7*'s record-breaking two weeks in space, longer than it would take to get to the Moon and back. The first rendezvous in space between *Gemini 6* and *Gemini 7* and the first docking in space with *Gemini 8* and its Agena rocket proved technologies that were needed for Apollo. When spacewalks proved to be more difficult than expected, Gemini astronauts perfected methods that allowed astronauts to work outside their spacecraft.

When Gemini first flew, the Soviet Union was ahead of the United States in the space race of the 1960s, with cosmonauts making longer and more sophisticated flights than were possible with Mercury. By the time Gemini ended in November 1966, the U.S. was firmly in first place. More importantly, the astronauts who would take Apollo to the Moon had gained invaluable experience flying in space, and the flight controllers in the new Mission Control Center in Houston, Texas, had learned important lessons that would prepare them for their work on Apollo. One of the toughest lessons came during *Gemini 8*, when a stuck thruster nearly caused the loss of the spacecraft and its crew, Neil Armstrong and Dave Scott. The two astronauts reacted coolly and promptly to the emergency, and flight director John Hodge, a British engineer who came to NASA from Avro Canada, ordered them home on the next orbit.

Chamberlin's colleague at Avro and NASA, Fred Matthews, called Chamberlin "the brains behind Gemini." James Rose, who helped to get Gemini running, said "there wouldn't have been a Gemini if it hadn't been for Jim Chamberlin." At the awards ceremony following the final Gemini flight, NASA awarded Chamberlin its Outstanding Scientific Achievement Medal for his work designing Gemini. "We've done a great deal more with Gemini than originally intended," NASA deputy administrator Robert Seamans said. "Gemini has done much more to open the way to the Moon than we could have hoped for five years ago."

Gemini's designers had plenty of input from astronaut Gus Grissom, and as a result, Gemini became the favourite spacecraft of the astronauts who had the chance to make a comparison. "Gemini was like flying a high-performance fighter," said astronaut Pete Conrad, who said it was more fun to fly than Apollo because Gemini operated through its control stick, while Apollo depended on its computers. As for Mercury, it could not maneuver or change orbits. The shuttle is also heavily dependent on computers. "I was amazed at my ability to maneuver," said Wally Schirra, who was the only astronaut to fly Mercury, Gemini and Apollo. "Gemini was magnificent to fly."

Although Chamberlin's Arrow didn't make it into production, many of the lessons he learned at Avro went into making Gemini the success it was. Even his biggest disappointment of the Gemini project, the paraglider, was eventually developed into a new type of aircraft that today populates the skies—the hangglider.

Between 1963 and 1970, Chamberlin remained with NASA as a troubleshooter for Gemini and Apollo. Some of the work he did on Apollo included helping fix design problems on the lunar module and joining the team that helped solve a set of mysterious failures on the Saturn V rocket during an Apollo test flight in 1968. Just before *Apollo 11* took the first astronauts to the surface of the Moon, Chamberlin chaired a group of engineers charged with making sure that the space suits and life support systems that Neil Armstrong and Buzz Aldrin wore during their historic moonwalk were ready to use.

Even before Apollo reached its lunar goal in 1969, NASA was looking ahead to space stations and a vehicle to carry astronauts into orbit and back—the space shuttle. Max Faget, then the head of engineering at

James A. Chamberlin

NASA in Houston, began design work on shuttles in the spring of 1969 and called on Chamberlin to head up a team that drew up designs for the shuttle. The project wasn't approved until 1972, and the final design was different from Faget and Chamberlin's ideas because of political tradeoffs with the air force, which wanted a larger shuttle with different flight characteristics.

By then, Chamberlin had left NASA and joined aerospace giant McDonnell Douglas. Although NASA selected another firm as prime contractor for the shuttle, McDonnell Douglas won a large portion of the detailed design work. Chamberlin was named the company's technical director for space shuttle engineering, and he continued in that job until he died of a heart attack on March 8, 1981.

Just five weeks later, the first shuttle flight lifted off from Cape Canaveral. NASA posthumously awarded Chamberlin its Exceptional Engineering Achievement Medal for his "engineering excellence" that contributed to the success of the shuttle. This medal was added to the medal he won for his work on Gemini and another he won for his contributions to the success of the Apollo Moon landings.

During the 1960s, Chamberlin became an American citizen, but he gained recognition in the Canadian media for his work on Gemini and Apollo. He had been virtually forgotten in his home country by the time of his death, but in 1996, Chamberlin was immortalized on a Canadian $20 coin along with the Avro Arrow. In 2001, Chamberlin was inducted into Canada's Aviation Hall of Fame. "His engineering genius, technical direction and leadership have been of significant benefit to Canada," his induction into the hall reads. "Further, his contributions to the United States space programs have given much credit to his home country of Canada."

Most Canadians who know of Chamberlin probably remember a character based on him in the 1997 made-for-television movie, *The Arrow*. The character, like much of the film, departs from the hard facts while remaining true to the spirit of the Arrow story. The film ends with a sequence in which an Arrow soars to the edge of space in an unauthorized flight after the cancellation, with a gleeful Chamberlin, sitting in the navigator's seat. While a flight has long been rumoured to have occurred, no solid proof has ever been found. And those who knew Chamberlin say he did not like to fly in aircraft he designed, rendering this story even more unbelievable. But the sequence provides a suitable salute to Chamberlin's role at NASA as a pioneer of space flight.

Even after more than four decades of human space exploration, the group of engineers who have designed spacecraft that carry humans is an exclusive one. In the U.S., Max Faget and Caldwell Johnson of NASA get most of the credit for designing Mercury, the Apollo command and service modules and the space shuttle. Tom Kelly and his team at Grumman Aircraft pulled together the Apollo lunar module. Jim Chamberlin was mainly responsible for the design of Gemini, and he left his mark on Mercury, Apollo and the shuttle. Chamberlin's contributions to human space flight also include the inspiration he gave one of his colleagues at both Avro and NASA who made vital contributions of his own to the Apollo lunar landings.

CHAPTER SEVEN

Owen Maynard and Apollo

ON JULY 20, 1969, THE WORLD WATCHED IN WONDER. Among the hundreds of millions who watched on television as *Apollo 11* astronaut Neil Armstrong made his historic first step on the Moon were those who came with Jim Chamberlin to NASA from Avro Canada 10 years earlier. Many of them stayed with NASA and played major roles in making this moment possible. Bryan Erb, an engineer from Calgary, managed the laboratory where the samples of lunar rock and soil would be tested once *Apollo 11* returned to Earth. He watched impatiently for Armstrong and fellow moonwalker Buzz Aldrin to pick up some rocks. "Pick up the damn rocks," Erb thought as he watched the ghostly figures on his TV screen. Others contemplated the enormity of the Moon landing. It had taken eight years, billions of dollars and more than 400,000 people to achieve. But the Canadian engineer who (next to Chamberlin) made the biggest contribution to the triumph of *Apollo 11* was sound asleep while Armstrong and Aldrin explored the Sea of Tranquility.

Owen Maynard spent the eight days of *Apollo 11* in a small room next to the Mission Control Center in Houston, Texas, where the spacecraft was controlled.

When he was on shift, Maynard had to be ready to deal with any problem that could imperil *Apollo 11* and its crew. Earlier on that memorable Sunday in July, Maynard closely monitored the action as Armstrong and Aldrin guided their lunar module, *Eagle*, to the lunar surface. The next day, Maynard was back at work as Armstrong and Aldrin lifted off the Moon and rejoined with Michael Collins on board the command module, *Columbia*. That maneuver was the rendezvous in space, 400,000 kilometres away from home, that had frightened so many people at NASA when the Apollo program got started.

But when Armstrong and Aldrin walked on the Moon, Maynard was off shift, so he went home and got the rest he needed for the stressful day ahead. Maynard's 18-year-old son Ross was "glued to the television" watching the moonwalk as his father slept, and he filled his father in on what had happened when he woke up early to head back to the space centre.

"At the time it impressed me that he had the discipline to know that he had to get a good night's sleep," Ross remembered years later. That discipline and determination were two qualities that had brought Owen Maynard to a top position in Apollo and kept him there through the program's ups and downs. Until the end of his life, Maynard never hesitated to remind people that he had no time for reflection or celebration until the flight was over and the astronauts were safely back on Earth. Maynard had pledged to carry out President Kennedy's lunar goal: "He didn't say go out and explore the Moon, or even bring back a piece of dirt." Maynard said. "He just said, land them and bring them back safely." Maynard helped see the crew of *Apollo 11* through their lunar liftoff, rendezvous and return to Earth. He promised to do the same four months later for *Apollo 12*, and when that mission was over, Maynard left NASA and returned to private industry. But in his decade at the agency, he made his own indelible mark on Apollo.

When the engineers from Avro arrived at NASA in April 1959, they were all assigned to help with the Mercury program, but Robert Gilruth and his team didn't want to lose any time preparing for flights that would follow Mercury. NASA formed an advanced projects group to look at new spacecraft and space stations, and among those chosen to join this group in 1960 were two Canadians, Owen Maynard and Bryan Erb. President Kennedy's lunar landing challenge was still a year in the future, but Gilruth directed the advanced projects group to start designing a new spacecraft, Apollo, that could fly three astronauts on long flights in Earth orbit and perhaps even into orbit around the Moon. Under the supervision of NASA's top designers, Max Faget and Caldwell Johnson, Maynard began work on early designs for what became the Apollo command and service modules. Erb, an expert on heat transfer, set to work designing Apollo's heat shield.

When President Kennedy directed Apollo to take astronauts to the Moon, Maynard looked on while people such as Faget, Johnson, Chamberlin and Gilruth debated how best to get there. Following Chamberlin's lead, Maynard was easily convinced that lunar orbit rendezvous was the best way to get to the Moon, rather than a direct flight with a single spacecraft. At Avro, Chamberlin had taught Maynard about the importance of modularity—splitting tasks into different areas so one machine doesn't have to do everything. "With the concept of modularity, you don't have all your eggs in one basket," Maynard said.

In the fall of 1961, Maynard began working on a vehicle that could take astronauts from the Apollo command module to the Moon's surface. Once Gilruth's group was convinced that lunar orbit rendezvous was the way to go, they had to convince others in NASA, including Wernher von Braun and his team in Huntsville, Alabama. Maynard was assigned to the team that went to Huntsville to get von Braun's approval. An important part of the presentation was a slide show that included Maynard's drawings of the lunar module.

Soon, the lunar orbit rendezvous plan and the lunar module (LM) design were given a "go." In the fall of 1962, the LM contract was given to the Grumman Aircraft Engineering Corporation in New York State. Work began on the new spacecraft. Thomas Kelly and his team at Grumman did much of the design work, but in the early days of the project, the man at NASA in charge of LM systems was Owen Maynard. During that time, many of the important decisions were made about the LM's design, including the number of legs on the descent stage and the design of the cabin. The engineers at Grumman and NASA had to decide what kind of footpads the LM would have, even though they didn't know exactly what the lunar surface was like. Some scientists feared that the Moon was covered with dust that would swallow up a spacecraft, but a robot vehicle that landed on the Moon in 1966 showed that the Apollo designers' educated guesses for the LM footpads were correct.

In 1964, Maynard was promoted to chief of systems engineering for Apollo. In that position, he reported to the program manager, and he was responsible for making sure that Apollo's myriad systems worked in harmony. Each of Apollo's components—the lunar module, command module and service module—had to work together, and the many systems inside each module had to work properly. In addition, the space-craft had to mate properly with its gigantic Saturn V launch vehicle, connect with NASA's worldwide communications network and control centre on Earth, and work with the launch systems and recovery forces. Maynard and his team had to make decisions about how to parcel out limited resources of space-craft weight, power and computing capacity.

❧◆❧

To deal with the pressures of a job with so much responsibility, Maynard needed to be strong. A com-pact, solid man with slicked-back brown hair and intense eyes, Maynard was born in Sarnia, Ontario, on October 27, 1924. His father was a postal clerk, and his mother was an amateur artist. In his boyhood, Maynard dreamed of building ships. He left school when World War II broke out and got a job building wooden boats for the navy. When he turned 18, Maynard enlisted in the Royal Canadian Air Force, and he graduated at the top of his class in flight training at Uplands, near Ottawa. Maynard finished just ahead of the son of a high-ranking RCAF officer, and since the classmate had made his desire to fly the de Havilland Mosquito well known, both he and Maynard were assigned to Mosquito training. The twin-engine, multi-purpose aircraft, built out of plywood, was a challenging aircraft to fly, and usually only veteran pilots were assigned to Mosquitos. This fact proved central to the lives of both men in different ways. The officer's son died in a training accident, and Flying Officer Maynard completed his training and went to Europe. But before he could fly in any combat missions, the war ended.

Maynard returned home, completed his high school and was accepted for studies in aeronautical engineering at the University of Toronto. When Maynard went looking for work to help pay for his education, his academic credentials did not impress a personnel officer at Avro Canada, but the fact that he had flown the Mosquito did. Maynard got the job. He financed his studies with work at Avro and another stint in the RCAF. When he graduated in 1951, Maynard got a permanent job at Avro. While he was with Avro, he pursued further studies in engineering and worked on the Arrow in several capacities, most notably on the aircraft's complicated landing gear.

Owen E. Maynard (centre, with headset) in the Spacecraft Analysis Room in Houston during the flight of *Apollo 11* in 1969. Thomas J. Kelly of Grumman Aircraft sits at left in the foreground.

During his university studies, Maynard married Helen, who was known as a great judge of people. Together, they raised a son and three daughters. As an engineer, Maynard made a point of knowing how things worked, and he loved to share his knowledge with his children and later his grandchildren. "If you wanted to know what time it was, he'd tell you how to make a watch," his daughter Merrill recalled.

When Bob Gilruth interviewed Maynard for a position at NASA, Maynard's experience flying the Mosquito again helped him win the job. When Gilruth asked what sort of work he was interested in, Maynard said, "I'd be interested in engineering or flying."

"Well, we aren't recruiting any flight crew," Gilruth replied, and Maynard was directed to a job in engineering. "I'm probably the first Canadian to volunteer to be an astronaut and the first to be rejected," Maynard remembered.

The cancellation of the Arrow project, which took Maynard, Chamberlin and Erb from Toronto to NASA and Houston, put about 2000 other engineers out of work. Some went to aerospace firms elsewhere in Canada or the U.S., some went to other engineering jobs and others left engineering altogether. Avro Canada was an example of how brain drain can work in both directions. Many of the skilled people who worked there had come from the United Kingdom to escape post-war cutbacks and retrenchment in British defence programs, and they were hoping to make new lives in Canada. Of the 31 former Avro engineers hired by NASA in 1959 and 1960, the majority had come from the UK. Only a dozen were Canadian. Prominent British members of the group included John Hodge, who became one of the first flight directors for NASA, along with legendary figures such as Chris Kraft and Gene Kranz; Rod Rose, who played a key role in drawing up flight plans for the Apollo missions; Peter Armitage, who worked on recovery systems for Mercury, Gemini and Apollo; Tecwyn Roberts, who became one of the pioneers of NASA's mission control room; trajectory expert Morris Jenkins; and Dennis Fielder and George Harris, who helped set up NASA's tracking network. The British members of what many people called the "NASA Canadians" even formed a NASA cricket team that played a handful of games in the early 1960s.

The Canadian members of the Avro group included people such as Maynard, Chamberlin and Bryan Erb, who was born and raised in Calgary, Alberta. Erb studied engineering at the University of Alberta and later in the UK. An expert on heat transfer, he was hired by Avro Canada to prepare the Arrow for the stresses it would experience in flight. After going to NASA and joining Maynard in working on the initial designs for Apollo, Erb became responsible for the engineers designing Apollo's heat shield that burned and carried away the heat as the Apollo command module struck the atmosphere and slowed down on its way to a splashdown in the ocean. Spacecraft in Earth orbit, such as Mercury, Gemini and the shuttle, first hit the atmosphere at a speed of 7.6 kilometres per second, but Apollo craft coming in from the Moon struck the atmosphere at 11 kilometres per second. These high speeds meant that designing and building Apollo's heat shields posed a whole new engineering challenge. The new heat shield was put to the test when *Apollo 8* orbited the Moon in December 1968 and brought its crew home. So, by the time that *Apollo 11* landed on the Moon, Erb had moved on to a new job managing the laboratory where Apollo's lunar samples were tested before being sent to scientists around the world.

Other Canadian members of Avro who worked on Apollo included Leonard Packham, a Saskatchewan native who helped design Apollo's communications systems; Fred Matthews, an engineer from Guelph, Ontario, who helped set up NASA's flight control centres; computer experts Stanley Cohn and Stanley Galezowski; structures expert Robert Vale; and heating expert George Watts. Richard Carley went from his groundbreaking design work on the Arrow's partially automatic fly-by-wire control system to building control systems for Gemini, Apollo and the shuttle. Eugene Duret joined other members of the group who worked on NASA's tracking network. During the Mercury orbital flights, starting with John Glenn's mission, Duret served as a capsule communicator in remote tracking stations.

More Canadians came to Houston and joined in the effort to put astronauts on the Moon. Two of NASA's top doctors treating the astronauts came from Canada. Dr. D. Owen Coons was a specialist in aviation in the RCAF who advised designers of the Avro Arrow on the human factors involved in the aircraft. The Hamilton native joined NASA in 1963 and worked as a flight surgeon for the Gemini and early Apollo missions, and he headed the NASA medical office at Houston. Dr. William Carpentier, who was born in Edmonton and raised on Vancouver Island, joined NASA after graduating from medical school. Dr. Carpentier was stationed on recovery ships during Gemini and Apollo missions to examine astronauts on their return and treat them if they were ill or injured.

And Canadians who stayed at home contributed to Apollo, too. Experts at the special projects and research division of de Havilland Canada, which later became

Spar Aerospace, built antennas for Mercury, Gemini and Apollo spacecraft. The legs and struts for the six lunar modules that landed on the Moon during Apollo were manufactured at Héroux Machine Parts Ltd., in the south shore Montréal suburb of Longueuil. The legs, which required precision milling, were left behind on the Moon with the LM descent stages and will stand there for many years into the future.

CHAPTER EIGHT

The Path to the Moon

IN MAYNARD'S TWO YEARS AS HEAD OF SYSTEMS ENGINEER-
ING under Apollo program manager Joseph Shea,
the Canadian and his colleagues were faced with
gigantic challenges as the Apollo spacecraft and the
systems that supported it came together. In 1965,
Maynard wrote a memo with a long list of major
problems facing Apollo. At the top came weight
growth, a problem that plagues almost every aircraft
and spacecraft program as systems grow in complex-
ity, adding more and more weight. The Saturn V
rocket would be able to boost only so much weight
to the Moon, and the weight of Apollo had grown
beyond projections "by a serious margin." The rocket
engines on the modules were providing less thrust
than expected in tests at the time, which meant less
weight could be moved, and serious issues remained
over the safety of both landing on the Moon and landing the command module once it returned to Earth.

In 1966, Maynard was transferred to become chief
of the mission operations division, a job that was
gaining urgency as Apollo moved closer to flight. One
of his jobs was to organize a meeting of the people
who were preparing for the first flight with astronauts

to the Moon. The Apollo Lunar Landing Symposium in June 1966 involved three days of detailed presentations that explained how the first landing on the Moon would take place. In the audience were leading people from NASA and the aerospace firms that were building spacecraft, rockets and infrastructure for Apollo. Maynard presented an overview of the mission, explaining that at various points of the flight, the mission could be changed if a system failed or the astronauts were in danger. These revised missions would ensure that the crew made it home safely.

The symposium reflected the growing confidence of NASA after the agency's recent successes. In the weeks preceding the symposium, the crew of *Gemini 9* had demonstrated rendezvous techniques that would be needed for Apollo, the *Surveyor 1* robot spacecraft had landed on the Moon, and a test article of the Saturn V rocket had been moved from the gigantic Vehicle Assembly Building at the Kennedy Space Center to the launch pad. And the first of crew of astronauts for Apollo was in training. Commander Virgil I. "Gus" Grissom and pilots Ed White and Roger Chaffee were preparing for a test flight of the command and service module.

As 1966 wore on, the astronauts and engineers in Apollo encountered a series of problems with the first spacecraft, including a balky environmental control system, but they pressed on with preparations for the launch of what would become known as *Apollo 1*, set for February 20, 1967. On January 27, Grissom, White and Chaffee were inside the spacecraft atop its Saturn IB rocket on the Florida launch pad for a rehearsal of the countdown for their flight. The test had gone on for hours, with unusual smells inside the spacecraft and problems communicating with the control centre.

The test was near its end, at 6:30 that evening, when the headsets of the test controllers were filled with cries from the astronauts: "Fire in the spacecraft!" *Apollo 1* was pressurized with pure oxygen, which is highly flammable, and the fire roared through sealed capsule until it cracked open. The astronauts had no chance to escape. Apollo's complicated hatch took more than 90 seconds to open, and the fire killed the crew in less than a minute.

NASA put Apollo on hold while the fire was investigated, and everyone involved in the program engaged in soul searching. Many people at NASA and at contractors involved in the fire left their jobs, including Joseph Shea and Robert Williams, the man who took over systems engineering from Maynard. As for Maynard himself, he never found out why no one made him share the blame for the fire, but he believes he was never blamed because he had always opposed using the hatch that was installed in Apollo and helped doom the astronauts. "There may have been a period when I may have been a potential guy to get blamed. Then they went back and remembered that they had gone against my guidance on the flight hatch."

While the precise cause of the fire was never found, NASA knew it had to redesign the Apollo capsule to eliminate fire hazards and make the hatch easier to open. Many other improvements were made, and Maynard said the pause after the fire "gave us a chance to hone up on what we were doing a lot better than we had before."

One day, George Low, the new Apollo program manager, dropped by Maynard's home and asked him to return to his old job in charge of systems engineering. "I will do this through the second lunar landing attempt," Maynard responded, "and whether or not we are successful, I will then resign and go back to industry." It was a promise Maynard kept.

Even before he returned to systems engineering, Maynard and his colleagues developed a series of Apollo flights leading up to the first landing attempt. The complex system had to be tested step-by-step before the full spacecraft was put to its ultimate test on a trip to the Moon. Maynard and his staff drew up what was known as the A to G sequence for Apollo. Various parts of Apollo—including the Saturn V rocket, the command and service modules, and the lunar module—were to be tested in a series of flights that culminated in the G flight, the mission that would take astronauts to the lunar surface and back to Earth.

The successful flight of the A mission using the Saturn V rocket and an unmanned spacecraft in November 1967 showed that Apollo was back on track. A repeat of the mission a few months later was not totally successful, but it pointed engineers to problems in the Saturn V that needed to be corrected before astronauts flew on board. The B mission tested a lunar module (LM) under ground control for a few hours in space. Then the Apollo team turned to the C mission, *Apollo 7*, that would carry astronauts Wally Schirra, Walter Cunningham and Donn Eisele in a shakedown cruise of the command and service module. Much to the relief of NASA, the flight, which took place in October 1968, was a total success.

Despite the success of its first test flight, the LM was still a long way from being ready to fly astronauts to the lunar surface. Maynard was busy running space-craft weight reduction programs and testing rocket engines that absolutely had to work if the astronauts were to survive the flight. The engineers slowly resolved the many issues plaguing the LM, but the problems meant that after *Apollo 7*, NASA would have to wait months before the LM would be ready for the D mis-sion—a test of the full Apollo spacecraft in Earth orbit.

Meanwhile, the Soviet Union was struggling with its own lunar program. Soviet cosmonauts had flown longer and higher than American astronauts until 1965, when the Americans took the lead with the first of 10 Gemini missions that weren't answered by a sin-gle Soviet cosmonaut flight. A few weeks after the Apollo fire, the Soviet Union launched cosmonaut Vladimir Komarov on board a new spacecraft type, the Soyuz, but Komarov died when his spacecraft's parachutes failed. By the summer of 1968, U.S. intel-ligence authorities received information that showed that the setback had not discouraged the Soviets, who were assembling a giant Moon rocket of their own. In the fall, they had Soyuz flying again, and a version of the craft had flown around the Moon. No cosmonauts were on board. Would the Soviets send cosmonauts on their next launch to the Moon?

When the news of the Soviet launches came in, George Low set Maynard and the other members of his team to work on a mission that was not foreseen in Maynard's A to G sequence. Instead of testing the LM, *Apollo 8* would fly a crew on board a command and service module into lunar orbit without the LM. When *Apollo 7* proved that the spacecraft was fit to fly, Low and the other leaders of NASA decided to send *Apollo 8* to the Moon. The astronauts and engineers worked overtime to prepare for the mission. They heaved a sigh of relief when no Soviet cosmonauts headed to the Moon when a launch window opened a few days before *Apollo 8*'s launch day. Years later, they found out that the Soviets held back because of lingering problems with their spacecraft that could have placed their cosmonauts' lives in danger.

On December 21, 1968, astronauts Frank Borman, James Lovell and Bill Anders became the first men to fly atop a Saturn V, putting them onto a path to the Moon. On Christmas Eve, *Apollo 8* reached the Moon and fired its engine to go into lunar orbit. On the journey out, the astronauts sent back televised images of the full Earth, and during their 10 orbits of the Moon, they televised close-up views of the Moon and took many dramatic photos, including an unforgettable shot of the Earth rising over the lunar surface. *Apollo 8*'s engine fired flawlessly to send the astronauts on their way home, and on December 27, the three astronauts returned safely to Earth from their historic journey. *Apollo 8* required hard work by people like Maynard and his team, who had to make sure that the space-craft, rocket, astronauts and controllers were ready for the ambitious flight. This work included planning the mission and preparing software for the computers on the ground and on board the spacecraft. *Apollo 8*'s successful re-entry at record-breaking speeds also proved that the heat shield designed by Bryan Erb's team worked properly.

A little more than two months later, *Apollo 9* lifted off on Maynard's D mission and proved the capability of the entire spacecraft, including the LM, in a series of tests in Earth orbit. By that time, NASA had decided against flying the E mission, in high Earth orbit. The next mission in Maynard's test sequence was the F mission, which would test the entire Apollo spacecraft, includ-ing the LM, in lunar orbit, without actually landing on the Moon.

Maynard concluded that *Apollo 8*, even though it didn't carry a LM, had fulfilled many of the goals of the F mission, and he suggested that the upcoming *Apollo 10* flight land on the Moon and carry out the final G mission. "I got convinced that once we had put the guys through the risk of the launch, and checking it out in Earth orbit, and going through the risk of solar events, and you do everything but land, I've already taken several years off the guy's life just sort of scaring him to death, and now I've got to, so to speak, take the same time off the lives of another set of guys to

Astronaut Buzz Aldrin stands on the Moon during *Apollo 11* on July 20, 1969.

actually accomplish the lunar landing objective," he remembered.

But many people believed that one more mission was needed before astronauts landed on the Moon, particularly the people who would be responsible for controlling the spacecraft. Led by legendary flight director Chris Kraft and strongly supported by Rod Rose, who had once worked with Maynard at Avro Canada, the controllers said they and the astronauts wanted to walk through everything but the landing to minimize the risks for the men making the first lunar landing attempt. The *Apollo 10* crew had been training for the F mission and would require new training to prepare for the actual landing. Maynard remembered that the argument ended when the *Apollo 10* commander, Tom Stafford, who would have become the first man on the Moon if his flight made the landing, decided that he favoured flying a mission without a landing. "So I accepted that and went on to do my best for *Apollo 10*."

The mission, which flew in May 1969, was a success and opened the door to a landing attempt on *Apollo 11*. Rod Rose recalled that the results of the flight proved the wisdom of flying the extra test before the landing, because *Apollo 10* brought back important information about how the spacecraft operated near the Moon and about unusual features of the Moon's gravitational field that affected the course of any spacecraft flying nearby.

A few weeks before *Apollo 11* attempted to fulfil Kennedy's national goal of landing astronauts on the Moon, Maynard was summoned to a meeting of top NASA managers and asked to take an extra hard and close look at whether everything was ready for the lunar landing—one of the parts of the mission that had not been attempted before. Maynard made a rare visit to the launch pad at the Kennedy Space Center and climbed inside the shroud protecting *Apollo 11*'s lunar module, *Eagle*, atop the Saturn V rocket. The bottom part of the LM, including the landing gear, was covered with thin mylar insulation, and Maynard wanted to make sure the mylar was properly attached and would not obstruct the opening of the landing gear or affect the landing radar. He took some tape

Apollo 11's command module, *Columbia,* is hoisted onto the deck of the USS *Hornet* following its splashdown in the South Pacific, July 24, 1969.

with him and tacked down the insulation in a couple of spots before concluding his inspection.

The endless meetings and problems were all finished on July 16, 1969, when Armstrong, Aldrin and Collins lifted off for the Moon. Four days later, Armstrong and Aldrin left Collins in lunar orbit aboard the command and service module, *Columbia,* and landed *Eagle* on the Moon. During their 22 hours at the landing site on the Sea of Tranquility, the two astronauts walked for 2.5 hours on the surface, picking up samples of lunar rock and soil, taking photos and leaving behind scientific experiments. During the flight, Maynard worked with managers from NASA and the companies that built Apollo to make sure that the ground controllers were ready for any emergency.

The morning after the Moonwalk, Maynard was back at his post, watching closely while the ascent stage of *Eagle* lifted off and then joined up four hours later with Collins in *Columbia*. This flight was the first to be attempted without the launch pad and techni- cians that usually accompany a launch, which was why Maynard was so intent on getting his rest instead of watching Armstrong's historic first step. After Arm- strong and Aldrin moved themselves and their lunar cargo into *Columbia*, the crew cast off *Eagle*. A few hours later, the three astronauts headed out of lunar orbit and back to Earth aboard *Columbia*, splashing down in the Pacific Ocean on July 24.

While Apollo was being prepared, public health authorities expressed concerns that returning Apollo astronauts could bring back germs from the Moon. Because of those fears, a three-week quarantine was arranged for the *Apollo 11* astronauts when they returned. Once they splashed down, the astronauts put on biological isolation garments that they wore until they were sealed into a special trailer on board the recovery ship. Dr. William Carpentier, the Canadian flight surgeon, and a technician joined the astronauts inside the trailer. Carpentier won the assignment because he was trained to leap out of helicopters in case the astronauts were in distress on splashdown. Carpentier found the astronauts in perfect health after their flight, and they stayed that way throughout the three weeks of their quarantine. The Moon was such a sterile place that there were no germs to catch, and later astronauts who visited the Moon were spared the quarantine. Strangely enough, Carpentier, who was exposed to lunar dust with the astronauts in quarantine, developed an allergy to the material. Back in Houston, Maynard and the others who helped make *Apollo 11* a success could finally celebrate.

After *Apollo 11*, six more expeditions were sent to the Moon, and all but *Apollo 13* made it safely to the Moon and back. The crew of *Apollo 13* survived their flight after an oxygen tank exploded on their way out to the Moon. Mission controllers devised ways to keep their spacecraft functioning until the astronauts could return to Earth.

The final three Apollo landings in 1971 and 1972 included lunar roving vehicles—a kind of car designed for the Moon—that carried astronauts miles away from their lunar modules for extended exploration of the area around their landing sites.

The Soviet Union tried four times to launch their giant N-1 Moon rocket, but it exploded on all four attempts. Once Apollo had beaten them, the Soviets never tried to land cosmonauts on the Moon.

The research carried out on the Moon during Apollo led scientists to a new theory about how the Moon was formed. Their theory has helped give humanity new ideas about the early history of the Earth. Scientists believe that an object the size of Mars struck Earth about four billion years ago. The debris released by the collision eventually formed the Moon. Scientists now believe that impacts from asteroids and comets play a major role in the history of the Earth, including an impact 65 million years ago that led to the extinction of the dinosaurs.

During the Apollo landings, Bryan Erb served as deputy manager and later manager of the laboratory at Houston that received the lunar samples and distributed them to scientists around the world, including geologists in Canada. Two Canadian geologists, David Strangway and Keith Richardson, worked in Houston where they prepared experiments for Apollo missions and helped train astronauts for their flights.

While they prepared to land on the Moon, Apollo astronauts visited several sites around the world where they might find geological formations similar to the lunar surface. In 1970, a group of astronauts that included the crew of *Apollo 15* visited the Canadian Forces testing ground in Suffield, Alberta, where they witnessed an explosion and then explored the crater that the blast created. The crews of *Apollo 16* and *Apollo 17* visited Sudbury, Ontario, to prepare for their flights. A large object struck the Earth near Sudbury 1.7 billion years ago, creating the area's nickel resources and other geological features like those that are found on the Moon.

Maynard left NASA after *Apollo 12*, the second lunar landing, and took a job in private industry with aerospace giant Raytheon in the Boston area. He worked on defence and space programs, including a plan to build satellites that could collect solar power and transmit it to Earth. In 1992, Maynard retired, and he and his wife Helen moved back to Canada. Maynard died in Waterloo, Ontario, in 2000. Like most of the Avro engineers who went to the U.S., he became an American citizen, but Maynard did not give up his Canadian citizenship and was proud of both countries.

Canada's most important contribution to Apollo was made by the engineers who went from Avro Canada to NASA. While everyone agrees that the Apollo astronauts would have made it to the Moon without the Avro engineers, the engineers from Canada showed that Canadians could work shoulder-to-shoulder with the best the world has to offer. As Bryan Erb said, "I think there was a feeling that people from Canada can do anything anybody else can do."

Jim Chamberlin played a crucial role at NASA in the 1960s, and so did Maynard, who was praised by Bob Gilruth, the director of the Johnson Space Center, for playing "a most significant role" in Apollo, including his early designs of the Apollo command module and lunar module. Gilruth said the Avro engineers were a "godsend" for NASA. And although Apollo was over, the space exploits of Canadians were just getting under way.

CHAPTER NINE

Communications Satellites

ALTHOUGH AMERICANS REVELLED IN THE TRIUMPH OF SEEING their astronauts walk on the Moon in 1969, and Canadians felt pride in the contributions the Avro engineers made to the feat, many people felt that the time had come to bring space programs back to Earth. As Apollo was soaring to the Moon in the late 1960s, people became more concerned about social issues on Earth, and they made it known that they wanted resources spent on alleviating those problems. Apollo came to an end in 1972, at the same time as the great post-war economic boom that made the spending for Apollo possible also ended.

Satellites were already changing lives on Earth by improving weather forecasts and opening new communications links, among other things. During the 1960s, satellite photos became a part of the daily weather forecast and saved countless lives by providing early warnings of hurricanes and other storms. And by the end of the 1960s, communications satellites made it possible for television pictures— including Neil Armstrong's first step on the Moon—to be broadcast live around the world and for telephone calls to be made almost anywhere. So while lunar

COMMUNICATIONS SATELLITES

exploration by astronauts was put on hold, satellites and space programs were becoming a permanent fact of life because they were paying real dividends.

The end of the 1960s was a period of transition for the American space program, and Canadian space activities also went through a major shift. Canada's fourth scientific satellite, *ISIS 2*, was being prepared for launch in 1971, after which the ionospheric research program would wind up. Canada was gearing up to launch its own communications satellites. That activity would be the top priority for Canada's space program in the 1970s. As part of the new emphasis, the Canadian government set up the Department of Communications in 1968 and made John Chapman assistant deputy minister with responsibility for satellite communications. The Defence Research Telecommunications Establishment at Shirley's Bay, where Chapman and his team built *Alouette 1*, was transferred to the new department and renamed the Communications Research Centre. The government also established Telesat Canada, a mixed private–public corporation, to operate the country's new communications satellites.

The Department of Communications was headed by a dynamic politician, academic and businessman from Montréal, Eric Keirans. He made the tough and unpopular decision to back Telesat's board when it decided to have Canada's first communications satellites built by an American contractor, Hughes Aircraft, rather than by the Montréal branch plant of the American communications giant RCA. Both Telesat and Keirans were anxious to get Canada's first communications satellite system running as soon as possible and within a reasonable budget, and RCA was not able to make firm promises on either. Keirans' decision was so controversial, even within the government, that he felt it necessary to go to Prime Minister Pierre Elliott Trudeau and threaten to resign if he could not carry out the decision to go with Hughes, which at the time was the world's top communications satellite builder. Keirans got his way, but he and Chapman remained anxious to build businesses in Canada that could profit from space industries. So when Hughes was chosen to build the first satellites, they agreed to use Canadian-built components not only in the satellites for Telesat but in their other satellites. Although Keirans' decision remained controversial, later events proved that he'd made the correct decision.

Hughes gave subcontracting work on the new satellites to a new company called Spar Aerospace, formed out of Northern Telecom and the division of de Havilland Canada that had worked on Alouette. In the fall of 1969, Telesat held a contest to name the satellites. A jury made up of communications theorist Marshall McLuhan, poet-songwriter Leonard Cohen and playwright Gratien Gélinas chose the name Anik, an Inuit word for "little brother." Canada's first communications satellite, *Anik A1*, was launched atop a Delta rocket from Cape Canaveral on November 9, 1972. A little more than a decade after launching its first satellite, Canada became the first country to launch its own domestic communications satellite into geosynchronous

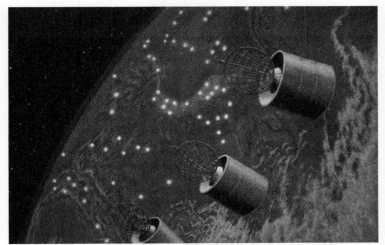

Artist's conception of the three Anik A communications satellites in space

~ⵙ⊃ℭⵡ~

orbit (the orbit 36,000 kilometres above the Earth used by most communications satellites). Because satellites in this orbit take 24 hours to make one circuit of the Earth, they appear to "hover" above the equator. Fixed antennas can only be used with satellites in geosynchronous orbit.

Chapman demonstrated his belief in Canadian space industry when he contracted out the work for *Alouette 2* in 1963 and when he wrote his 1967 report that called for such an industry. With the three Anik A satellites being launched in the early 1970s, satisfying the government's priority to provide radio, television and telephone communications to every part of the country, the Canadian government and

contractors could turn to developing the space industry. The subcontracting work with Hughes was a start,
and the Communications Department supported
research on communications satellite technologies at
its Communications Research Centre. The department
also built a large facility at Shirley's Bay to assemble
and test large communications satellites and other
spacecraft. Named the David Florida Laboratory after
the late manager of the ISIS program, the facility
opened for business in 1972.

These steps laid the foundation for Canada's communications satellite industry, but Chapman and his
colleagues wanted to make Canada a leader in satellite technology. They cancelled a third ISIS satellite
and divert the funds to a new project that would result
in the first of a fresh generation of communications
satellites.

The first generation of Canadian communications
satellites broadcast in C-band, which requires limited
power in the satellite but larger antennas, usually the
three-metre-wide dishes that became familiar sights in
the late 1970s and early 1980s. The new satellite would
broadcast in the Ku-band, which requires more powerful satellites but smaller antennas, such as the 0.6-metre
direct-to-home satellite antennas in wide use today.

In 1971, Canada launched the Communications
Technology Satellite (CTS) program. Work on the satellite began in 1971 after a study led by Chapman.
Like the Alouette and ISIS programs, the CTS project
started out as a cooperative program between Canada

and NASA. Soon the European Space Agency was on board, marking the beginning of the cooperation between Canada and Europe in space that continues to this day. The project manager for the first Canadian-built communications satellite was Colin Franklin, who played an important role in Alouette. RCA Canada, which became a division of Spar during the life of the project, built the satellite, with assistance from Spar and other Canadian aerospace firms.

The Communications Technology Satellite, renamed *Hermes* after its launch on January 17, 1976, was the first communications satellite to broadcast in Ku-band. In the spring of 1977, it became the first to make a direct-to-home satellite television broadcast when it beamed a Stanley Cup hockey game to the home of the Canadian trade counsellor in Lima, Peru. *Hermes* operated for nearly four years, twice its planned lifetime, and during that period it pioneered the broadcasting of educational programs, the linking of doctors, patients and families between cities and isolated communities, teleconferencing and technologies that today allow television news crews to uplink their reports to television stations via satellite.

Before *Hermes*, communications satellites were drum-shaped vehicles that spun to stabilize themselves. *Hermes* was the first to use three-axis stabilization that allows larger antennas and more solar panels to collect power. That design is now the standard in today's communications satellites. *Hermes* was the most powerful communications satellite built at the time, and its launch and deployment sequence was far more complex than any previous communications satellite because it carried large antennas and solar panels. The groundbreaking work of *Hermes* was followed up by the *Anik B* satellite, launched in 1978 and broadcasting in both C-band and Ku-band.

Today, many communications satellites broadcast in Ku-band, and direct-to-home satellite is a growing industry. Satellite television services such as Bell ExpressVu and Star Choice are gaining popularity, as is satellite radio. Canada's significant contribution made these services possible. Telesat is a major provider of direct-to-home services with its Nimiq and late-generation Anik satellites.

Chapman did not give up on his dream of making a Canadian company into the prime contractor for communications satellites. In May 1979, shortly before Chapman died, Telesat contracted with Spar to build two Anik D satellites, which were launched in 1982 and 1984. Spar also built the more advanced Anik E satellites, which were launched in 1991, and the MSAT mobile communications satellite. Spar also won foreign business when it won the contract to build two Brasilsat communications satellites for Brazil in the early 1980s. In the 1990s, Spar found the going tough as large aerospace firms began merging. Boeing, for example, swallowed Hughes, and European satellite manufacturers also merged. To add to the challenge, larger communications satellites were being built, which meant that fewer were needed. In 1998, Spar sold its satellite division, and the company left the space business completely the next year.

While operating communications satellites, Telesat and Spar developed a base of experience in operating satellites and receiving and transmitting stations, and they have been able to sell their expertise. But some of this knowledge was hard won. A few days after *Anik E2* was launched on April 4, 1991, it arrived in geosynchronous orbit and began deploying its separate Ku-band and C-band antennas. The C-band antenna failed to deploy, and the stuck antenna in turn blocked important sensors used to point the satellite. Working with limited information and using *Anik E1*, which was still on the ground at the David Florida Laboratory, as a test article, Canadian controllers were able to determine that insulation blankets had prevented the antenna from deploying. They also concluded that they could free the stuck antenna if they spun the satellite, a complicated and difficult maneuver. The antenna failed to spring free the first time *Anik E2* was spun. When they tried a different type of spin on July 2, the antenna sprung free, and the satellite began operating properly. When *Anik E1* was launched later in the year with changes to its thermal blankets, its antennas deployed properly.

But the problems weren't over. On January 20, 1994, Canadians lost a number of broadcast services, including many cable TV channels, when both Anik E satellites failed within 70 minutes of each other. It turned out that a magnetic disturbance had caused both satellites' stabilization systems to fail—a computer chip was susceptible to the magnetic storm. The disturbance had also struck another communications

Artist's conception of an Anik E satellite in space

~∞X∞~

satellite a few hours before the two Aniks. *Anik E1* was brought back online within hours, but *Anik E2*, true to its troubled heritage, had more problems. It was eventually stabilized using its thrusters, but this solution took months to implement and cut short the satellite's useful service life. Similar incidents have struck other communications satellites. Operators now take care to prevent problems from solar storms, which can shower satellites in geosynchronous orbits with radiation.

In spite of occasional business and operational problems, Canada's satellite communications industry continues to thrive. Another part of the business involves building satellite stations and ground systems. SED Systems of Saskatoon, Saskatchewan, and MDA (MacDonald Dettwiler and Associates) of Richmond, BC, have thrived in the business. MDA has grown into Canada's largest space contractor, a status confirmed late in 2005 when it acquired the former RCA and Spar satellite operations in Montréal. Com Dev International of Cambridge, Ontario, which began in 1974 and has since built components for more than 500 satellites, has also become a major player in the space industry. Telesat became a private corporation in 1992, and in its history it has launched 14 Anik and two Nimiq satellites into orbit, all successfully.

The Canadian government continues to support the satellite industry with research at the Communications Research Centre and satellite integration at the David Florida Laboratory at Shirley's Bay. This work, which is supported by the Canadian Space Agency and Industry Canada, continues to take up a major part of Canada's space budget, but the spending has paid off in large-scale exports that generate high-technology jobs across Canada. The Canadian communications satellite industry on its own earns more than $1.8 billion, much of it in exports to other countries, and accounts for three-quarters of Canada's space industry revenues.

NASA launched the first weather satellite, *TIROS 1*, on April 1, 1960. The first photograph it sent back showed clear skies over Nova Scotia and New Brunswick and clouds over New England. Canada quickly moved to directly receive photos and other data from *TIROS 1* and the weather satellites that followed.

Canada continues to rely on American weather satellites, but Canada is involved in another lifesaving service based on these satellites. In 1979, Canada joined the U.S., France and the Soviet Union to establish a system that uses space-borne receivers on satellites to pick up emergency search and rescue (SAR) signals from ships and aircraft. The system got underway when the first such receiver was launched on the Soviet *Cospas* satellite in 1982. It was used for the first time on September 9, 1982, when signals from a downed aircraft in British Columbia were picked up. Three people were rescued. Today the Cospas-Sarsat system, which is headquartered in Montréal, has expanded to include many nations. Receivers are carried on weather satellites and other satellites that regularly fly over all parts of the globe. By the end of 2004, the Cospas-Sarsat system had been involved in the rescues of more than 18,800 persons in more than 5300 incidents.

Weather satellites marked the beginning of remote sensing from space. By the time of the Chapman report, NASA was working on satellites that could search for mineral resources from space and keep track of changes, both natural and human-related, on Earth. Chapman's report raised the importance of remote sensing for Canada, as did a report from the Science Council of Canada. In 1969, the federal government set up a special committee on the subject.

By the time NASA launched its first Landsat remote sensing satellite in 1972, Canada had its own Centre for Remote Sensing to coordinate Canada's involvement with Landsat and later remote sensing satellites, such as the U.S. *Seasat* oceanographic satellite and European satellites, such as *SPOT* and *Envisat*.

Chapman and others were thinking about a Canadian remote sensing satellite. By 1977, Chapman's interdepartmental committee on space persuaded the federal government that Canada should build a remote sensing satellite that used radar. A few years later, the government gave the go-ahead to Radarsat. Spar Aerospace won the contract to build *Radarsat 1*. The sophisticated spacecraft was finally launched on November 4, 1995. Radarsat uses a sophisticated form of radar, called synthetic aperture radar, that is used to create computer images based on radio waves sent from the satellite and reflected back from Earth. Unlike other satellites, Radarsat can image the ground and the ocean surface at any time of day without being affected in any way by clouds.

These images are useful for mapping, but that is just the beginning of the applications for Radarsat images. They are also used to keep track of the movement of icebergs and other changing natural phenomena, such as changes in rivers. Radarsat images were used in 1997 to keep track of the Red River when it flooded much of southern Manitoba and parts of the U.S. Radarsat information has also been used to track illegal fishing and dumping of oil in the oceans. Other uses include tracking of farming and urban land development,

Artist's conception of *Radarsat 1* in orbit

~✖~

monitoring security, searching for oil reserves, finding
other natural resources and mapping wetlands. The
satellite is also used to help Canada to protect its
sovereignty in the Arctic. Besides being used by
Canadian governments, images from Radarsat are
sold on the international market by MDA Geospatial
Services, a branch of Canadian space contractor
MDA.

MDA was heavily involved in the fabrication of the
first Radarsat, and it won the contract to build a new
and more powerful satellite, *Radarsat 2*, due for launch
in 2007. The satellite offers much higher resolution
than the original *Radarsat 1*, and it will be the most
sophisticated satellite of its kind. Although *Radarsat 1*
was built and launched in cooperation with NASA,
the U.S. agency declined to work with Canada on

Radarsat 2, reportedly because of security concerns about the high-resolution images available from the satellite. Although Canada has reached an agreement with the U.S. government about its security concerns, prime contractor MDA is working with a European contractor on *Radarsat 2*.

The Canadian Space Agency has adopted a "niche strategy" to develop Canada's space program. This means concentrating on areas of expertise for Canada, such as communications satellites, receiving stations and the Radarsat remote sensing satellites. Canadian space industry has also moved into building equipment for the Americans' Global Positioning System (GPS) satellites, which allow people to navigate and keep track of where they are and where they are going.

Although communications, remote sensing, navigation, receiving systems and satellite components form the bread-and-butter of Canada's space industry, the Canadian space program found a practical means of getting involved in the best-known part of the space industry—sending astronauts into space.

CHAPTER TEN

The Canadarm

CANADA'S SPACE PROGRAM HAS A VERY PRACTICAL ORIENTA-
TION. So it's not surprising that the main reason Canada has astronauts today is that the U.S. space shuttle
became part of Canada's strategy to develop high-tech
industry. Back in 1967, when John Chapman and his
colleagues prepared their report on Canada's space
program, they were silent about sending Canadians
into space. At the time, the U.S. and the Soviet Union
were locked in their race to the Moon, and there
didn't appear to be any room for other countries to get
involved. Two years later, shortly after the *Apollo 11*
astronauts landed on the Moon, NASA officials came
to Ottawa to see if Canada was interested in getting
involved in NASA's plans for flights after Apollo. The
problem was, NASA was talking about ambitious
projects such as space stations, shuttlecraft and expeditions to Mars. The U.S. government had approved
none of these plans, and there were questions about
whether the projects had any support at all. The
Canadian government said no.

Three years later, the picture was changing dramat-
ically. Apollo was winding down. Early in the year,
President Richard Nixon approved NASA's plans to

build a space shuttle that the agency hoped would reduce the cost of going into space. NASA's plans for space stations and trips to Mars were gone.

One day in the early 1970s, a Canadian engineer named Lloyd Secord made a sales call to NASA. Was the agency interested in a space telescope his engineering firm was designing? The answer was a polite no, but during the conversation, Secord mentioned that he was working on a robot arm for use inside nuclear reactors. His contact at NASA mentioned that the space agency wanted to build a large robot arm to move satellites and equipment around the shuttle's payload bay. Thus begins the story of what NASA calls the shuttle Remote Manipulator System (RMS), and Canadians call the Canadarm.

When Secord got home, he got in touch with other firms that might be interested in such a project, including Spar Aerospace. Soon the discussions included government officials such as Frank Thurston, the head of the National Aeronautical Establishment of the National Research Council of Canada. Thurston became an enthusiastic backer of the idea, and he soon won support from John Chapman and from the science minister at the time, Jeanne Sauvé (who in the 1980s became Governor General of Canada). In 1974, the Canadian government proposed to NASA that Canada build the robot arm for the shuttle.

In 1975, the Canadian government agreed to pay the $100 million cost to develop and fabricate the first RMS robot arm for the shuttle. NASA agreed that it would purchase three more arms from the Canadian contractors. The project began under of the National Research Council (NRC), with Spar as the main contractor and the Canadian flight simulator firm CAE Industries, Secord's firm DSMA Atcon and RCA Canada as the major subcontractors.

Over the next six years, building the arm presented several challenges to the team of 800 people from both the NRC and the contractors, headed first by Garry Lindberg and then Art Hunter from NRC and John MacNaughton from Spar. A major hurdle was devising a computer control system for the arm. Such a system had never been built, and NASA imposed rigid quality standards on everything flying on the shuttle. NASA expected the arm to move precisely to where it was directed and required that the arm be designed so that it would still function even with a major failure. If the arm experienced two major failures, NASA stipulated that the arm could not pose a danger to the shuttle or crew.

An important part of this effort was building an effective simulator facility, which Spar constructed in the Toronto area. The facility was not only used to train astronauts but also to check out design changes in the arm before they were implemented. The simulator's original computer software was known as ASAD, but unlike most acronyms, this one was whimsical—it stood for "all singing, all dancing"— because the simulations dealt with all kinds of challenges. The arm itself could not be fully tested until it was in space because it was too light to lift its own weight on the ground, but limited testing was

Painting by space artist Paul Fjeld of the space shuttle with an extended Canadarm

done with the arm mounted on air-bearing supports in a room with a specially smoothed floor.

The arm itself would be exposed to the harsh vacuum of space with its temperature extremes. The gears for the arm were precision machined from a special stainless steel alloy that had been heat treated, but even that material would shrink under space conditions, so the gears had to be machined to a larger size than needed, then heat treated until they shrunk to the proper size.

An NRC publication described the arm this way: "Canadarm is a robot analogue of the human arm—its nerves of copper, its bones of graphite-fibre-synthetic tubes and its muscles of electric motors. Each of these motors is no larger than a telephone handset, and works on direct current. Serviced by gearboxes with gear ratios in the order of 1800:1, the motors lie buried in Canadarm's metal joints. Unloaded, the arm can move its tip at about 70 [centimetres per second]. This diminishes to 5 [centimetres per second] under its maximum load-carrying capability of almost 30 tonnes." While objects are weightless in space, they still have mass, which comes into play when they are being moved.

The arm is 15.2 metres long, 38 centimetres in diameter and weighs 410 kilograms. It has a shoulder joint with two degrees of movement, an elbow joint with one degree of movement and a wrist joint with three degrees of movement. At its end is a device called an end effector that resembles an empty shirt cuff. Satellites and other payloads handled by the Canadarm are fitted with a knob called a grapple fixture. When astronauts want to grab a payload, they move the end effector so that it encloses the grapple fixture. Then the astronaut pulls a trigger that causes three wires inside the end effector to snare the grapple fixture and pull it inside the end effector until it is firmly grappled. The grapple fixture looks like a simple device, but it required precision engineering. NASA insisted that astronauts must be able to manually detach grapple fixtures from payloads in case the Canadarm fails while it is holding the payload.

Astronauts control the Canadarm from a console in the back of the shuttle's flight deck, next to the windows facing into the payload bay where the arm is located. The arm can be moved in two automatic modes, or it can be manipulated using two control sticks, either joint by joint in case the shuttle computer fails or semi-automatically. The astronauts can not only see the arm through shuttle windows, but they can also see what they're doing with the help of television cameras mounted on the arm's elbow and wrist.

Before NASA agreed to have Canada build the arm, the NRC and its contractors underwent rigorous scrutiny from the American space agency. Another round of hard examinations came in 1978, when teams of NASA engineers came to Toronto to conduct a two-week critical design review with 100 members of the Canadian RMS team. The entire program was reviewed and any remaining problems were discussed. One of the concerns was the possibility that the arm could lose control. The Canadian team quickly developed a computer program that ensured that a runaway arm could only run away a couple of feet.

Finally, in February 1981, the first Canadarm was ready for delivery. The arm was handed over at a special ceremony at the Spar plant, and it was trucked to the Kennedy Space Center in Florida in a high-security convoy.

While the Canadians were building the RMS, NASA was preparing five shuttles during the 1970s. In 1977, the shuttle *Enterprise* made a series of flights off the back of a modified Boeing 747 ferry aircraft to test its landing characteristics. When it returns to Earth, the shuttle is a glider (with no engines). NASA wanted to make sure that the astronauts could land it safely, since they would have no second chances. *Enterprise* was not equipped to fly in space, so it was used to prepare the launch pads at Cape Canaveral for shuttle operations. The first shuttle to fly into space was *Columbia*. This shuttle went through several difficulties as NASA and its contractors prepared it for flight, but finally, on April 12, 1981, *Columbia* left its launch pad with astronauts John Young and Robert Crippen on board. The shuttle flew according to plan and landed safely in California two days later.

The Canadarm was scheduled to make its first flight on *Columbia*'s second mission, in November, but a few weeks before the launch, NASA considered removing the arm from the flight. *Columbia* had encountered more severe vibrations than expected at its first launch, and NASA was afraid that the arm would be damaged and could threaten the mission. After engineers installed a new system on the launch pad to reduce launch vibration, NASA decided to press ahead with testing the arm.

But there was one more change. When the arm was enclosed in its white insulation blankets, the arm sported a Canada wordmark (the word "Canada" with a small Canadian flag). Art Hunter and his assistant manager, Bruce Aikenhead, saw that the European Spacelab module was going to sport a European Space Agency logo, so they sought and gained approval for the Canada wordmark on the arm. When the arm was installed inside *Columbia*'s payload bay at Cape Canaveral, officials hurriedly arranged to have an American flag displayed in the cargo bay as well.

Columbia's first flight with the Canadarm was delayed several times, but finally it began on November 12, with astronauts Joseph Engle and Richard Truly on board. Like many other astronauts, Truly had come to Toronto to train on the Canadarm. Soon after, *Columbia* became the first shuttle to make a return flight to space. But one of its three fuel cells started to act up, so NASA decided to cut the flight in half from five days to 54 hours. The astronauts would have only four hours to test the arm rather than the planned 12-hour trial.

Truly moved to his station on the second day of flight, on Friday, November 13, 1981, and unberthed the arm while the Canadians at mission control in Houston held their collective breath. The Canadarm performed nearly perfectly. *Columbia* beamed back television pictures of the arm, bearing its Canada wordmark, with the Earth in the background. The television cameras on the arm captured spectacular views of the shuttle cabin and a shot of Truly holding a sign in a window that read, "Hi Mom." A joint failed to move because of a broken wire, and a television camera broke down. But as Truly later said, it "performed magnificently throughout the test." Added the NRC's Hunter, "We're obviously delighted that it went so well." The flight ended safely. On its next flight in March 1982, the Canadarm moved its first load around the payload bay. After *Columbia*'s fourth test flight in July, the shuttle and the Canadarm were ready.

Over the next three years, the shuttle fleet grew with the addition of orbiters *Challenger*, *Discovery* and *Atlantis*. Spar delivered three more Canadarms to

NASA for use with the new spacecraft. During 20 flights that followed the four test flights, the shuttles deployed a large number of satellites, including four Aniks. The shuttles were also used for several dramatic rescues of broken satellites. The Canadarm played a key role in all of those missions. A 1984 flight of *Challenger* was launched to rescue the *Solar Max* satellite, and NASA planned to use two astronauts with jet backpacks to do the job. But the astronauts could not grapple the wayward satellite. Finally, the Canadarm was called into action. The arm grabbed the tumbling satellite on its second attempt, and the crew of *Challenger* repaired *Solar Max* and sent it on its way.

When *Challenger*, its crew and its Canadarm were lost on the 25th space shuttle launch on January 28, 1986, NASA ordered a replacement shuttle, *Endeavour*, and a replacement arm. Since then, the arm has played a key role in shuttle missions, including the exacting repair and refurbishment mission for the *Hubble Space Telescope* in 1993 and more recent missions building the International Space Station. And the arm has functioned without ever having a failure that affected a mission.

Today, the Canadarm team works for the Canadian Space Agency and MD Robotics, the successor organizations to the space division of NRC and Spar Aerospace. But work on the arm continues. NASA arranged with MD Robotics to upgrade the arm's joints and its software so that the RMS can move heavier loads. And the success of the Canadarm has resulted in sales of specialty equipment built for the shuttle arm and special robot arms for use in nuclear reactors. In its first 20 years, the Canadarm generated an estimated $700 million in export sales for Canada, and today that figure is much larger.

Canada has always given its space program a practical orientation, and that includes its participation in the space shuttle program. The Canadarm would lead to many things in space, including opening the door for Canadians to join the ranks of the astronauts.

CHAPTER ELEVEN

Marc Garneau: Canada's First Astronaut

DURING THE TWO HOURS MARC GARNEAU SPENT IN HIS SEAT aboard the space shuttle *Challenger* waiting for launch early in the morning of October 5, 1984, he no doubt had time to reflect about many things. Probably one of them was how in a little over 10 months he had moved from his post as a naval officer in the Canadian Forces to the brink of becoming the first Canadian to go into space. A few months before he became an astronaut, Garneau had no idea that it would be possible for any Canadian, let alone him, to climb aboard a space shuttle. But for Garneau, fate put him on board the shuttle in an unusually short time.

Indeed, most Canadians spent the space age watching American astronauts and Soviet cosmonauts fly into space without knowing when any Canadians might get a chance to join them. That began to change in 1979, when NASA sent a letter to the Canadian government inquiring whether Canada wanted to set up its own astronaut program. A number of developments were behind the letter. The year before, America's Cold War adversary, the Soviet Union, began launching "guest" cosmonauts from friendly countries on short trips to their Salyut 6 space station. The first space traveller who was neither American nor Soviet was Vladimir Remek of Czechoslovakia. The propaganda impact of these flights was not lost on NASA.

The American space agency was building its space shuttle fleet that year, and Canada was helping by building the Canadarm. At the same time, the European Space Agency was building Spacelab, a scientific module that would fly in the shuttle's payload bay and allow astronauts to perform a variety of experiments. In 1978, the ESA began training its first astronauts to fly aboard the shuttle and Spacelab. Since Canada was making such a large contribution to the shuttle program, it made sense that Canada also be invited to fly its own astronauts. So NASA wrote Dr. John Chapman, the head of Canada's Interdepartmental Committee on space. The letter went unanswered because Chapman died that fall, and Canada's space policies were in question because of an upcoming federal election.

In 1981, the Americans again informally asked Canadian officials if the country wanted to fly astronauts. In the fall of 1982, at a celebration marking 20 years since the launch of *Alouette 1*, a NASA representative made the invitation formal. Staff from the National Research Council of Canada considered how to respond, including looking at how many astronauts Canada would need and what they could do in space.

On June 8, 1983, the space shuttle *Enterprise* made its only visit to Canada on the back of its Boeing 747 carrier aircraft. When the aircraft touched down at Ottawa International Airport with top NASA officials on board, Canada's science minister, Don Johnston, announced the official launch of the Canadian Astronaut Program. A few weeks later, newspapers carried an out-of-this-world ad looking for astronauts. "Candidates should have a university degree and experience in system development, integration and operation, or in vestibular physiology and/or motion sickness," the advertisement said. "Candidates must be able to meet appropriate medical requirements. Practical experience in flying would be an asset, as would knowledge of both official languages." Garneau noticed the ad while reading the newspaper at home in Ottawa. Soon he was among the 4380 people who sent in an application before the August 8 deadline.

The job of selecting Canada's first astronauts was left to the National Research Council, which was put in charge of the astronaut program. Many of the applications came from people who weren't qualified, and a few even came from children. "We had an awful lot of unsuitable people," one official said. "I'd say more than 50 or 60 percent. We had some strange people." But many resumés came from highly qualified applicants. By the end of August, the NRC had winnowed down the 4380 hopefuls to 1800. The NRC wrote to them asking for more information about their work experience, education and health status. The responses resulted in a further paper selection process that left only 68 candidates.

A five-member selection committee went to Ottawa, Montréal, Toronto, Calgary and Vancouver to interview the 68 hopefuls in October and November. Unlike

American astronaut selections, where the identities of candidates are kept secret until those selected are revealed to the world, the media was allowed to interview candidates when the selection committee came to their city.

At one of the media events, Karl Doetsch, the director of the Canadian Astronaut Program, suggested that the media availability was one of the tests the applicants had to pass in addition to their formal interviews. "It's important that the human element of the endeavour be spread to the public because they are paying the bills."

The candidates the media met were highly trained engineers, physicists, medical doctors or physiologists. One of them, John Steeves, a University of British Columbia neurobiology professor, told the Vancouver press conference that he and his fellow applicants were different from the daring 1950s test pilots who became the first American astronauts.

"We happen to be chosen for what we're doing right now," said Steeves. While he was interested in space travel, he did not point his career in that direction. And like many of those who didn't make the final cut, Steeves has gone on to a distinguished academic career, making a name for himself with his research on repairing damaged spinal cords.

The selection committee reduced the 68 applicants to 19 candidates, who were invited to Ottawa later in November for a round of medical tests and further interviews. Because their names were already public, the finalists, who included three francophones and

Canada's first astronaut team, selected in 1983. Left to Right: Ken Money, Robert Thirsk, Marc Garneau, Steve MacLean, Roberta Bondar, Bjarni Tryggvason.

one woman, Ontario physician Roberta Bondar, had to deal with media interest while they went through the final process. Finally on December 3, a Saturday, Ray Dolan, who was in charge of personnel for the NRC, phoned all 19. Six of them got good news—Marc Garneau; Roberta Bondar; Ken Money, a physiologist who was a world authority on motion sickness in space; Steve MacLean, a physicist who was already working on a computerized space vision system; Bob Thirsk, a physician and engineer; and Bjarni Tryggvason, an engineer and physicist. For the six, the phone calls led to celebrations. MacLean, who was also a gymnast, did a back flip at the end of his telephone call.

Two days later, Canada's first group of astronauts was revealed at a press conference in Ottawa. At the time of the announcement, NASA had promised only two shuttle flights for Canadians, one in November 1986 to test the Space Vision System, a Canadian-built system that allows the shuttle and the Canadarm to automatically locate and join up with satellites, and another flight in 1986 to carry out research on motion sickness, which affects many astronauts.

Over the next few weeks, the astronaut trainees wrapped up their previous jobs, and if necessary, moved to Ottawa. Garneau already lived there, but he had to shift his workplace to the astronaut program's nondescript office in an obscure building on the NRC's campus on the eastern fringe of Ottawa. The astronauts barely had time to find their desks in late January 1984 when they were summoned by the president of the NRC, Larkin Kerwin. He told them that NASA had offered to fly one of them a year ahead of schedule, in October. After brief consideration, the NRC decided that it could prepare experiments to justify such a flight and had agreed to NASA's offer. Not only would the selected astronaut have to scramble to prepare for the flight, but NRC personnel would also be busy preparing their experiments.

The six astronauts began training at two Canadian space medicine laboratories, one in Toronto run by Money and another at McGill University in Montréal run by physiologist Douglas Watt. At Watt's lab, they tried a "sled" that Watt hoped to send on board to test for space sickness. At Money's lab at the Defence and

Civil Institute of Environmental Medicine, they were spun in a device designed to induce motion sickness. Since Money had already taken many runs on the device, he didn't try it, but the other five did. None threw up, although all experienced motion sickness. They also tried out the altitude chamber at the institute to prepare for later training in aircraft operating at different air pressures. Soon their director, Karl Doetsch, had to decide who would fly in October based on the observations he made during the two months the astronauts were together. On March 13, he called a meeting and told them that Garneau would get the flight, with Thirsk as his backup. The next day, the news was made public.

Garneau, the public learned, had been born on February 23, 1949, in Québec City to a French Canadian father and an English Canadian mother. He was fluently bilingual. He was the second of four boys. Since his father, Andre, was in the military, the Garneaus moved several times during Marc's childhood, including postings in Germany and England. After attending what his mother Jean called "12 or 13 schools," the young Garneau went to the Royal Military College of Canada at Kingston followed by studies at the Imperial College of Science and Technology in London, where he earned a Ph.D. in electrical engineering.

Garneau was a combat systems engineer aboard HMCS *Algonquin* from 1974 to 1976. While serving as an instructor in naval weapon systems, he designed a simulator to train weapons officers. At the time he was selected as an astronaut, Garneau was a commander in the Canadian navy. His expertise in naval combat systems led him to a job in Ottawa, where he was responsible for approving the design of new naval weapons systems. He and his wife Jacqueline had eight-year-old twins, Simone and Yves. Garneau's hobbies included jogging, swimming, scuba diving, playing squash, making wine and working on cars, a list that fit in well with those of many American astronauts. While Garneau is self-controlled in public, his friends say he is a funny and patriotic man.

While the reasons Garneau was selected were never made public, many noted that he was the only astronaut who was effortlessly bilingual. As Garneau would later learn, communication in both languages was an important part of the Canadian astronauts' job. Even before Garneau's selection, the other astronauts were hard at work on language lessons.

After the announcement, the astronauts' training resumed. At first, Garneau and Thirsk continued working with their four colleagues in Canada, doing further work with the motion sickness simulators and going to suburban Toronto to visit the Spar Aerospace plant where the Canadarm was made. There, they tried out the simulator where American astronauts came to learn how to operate the Canadarm. Garneau and the other Canadian astronauts would not use the arm in their scheduled flights, but they needed to know how it operated. And like all other astronauts, the Canadians spent hours attending lectures and poring over books to learn more about the space shuttle and the environment of space. In the spring,

Canadian astronauts Marc Garneau and Robert Thirsk during training for
Garneau's first flight into space in 1984. Thirsk served as Garneau's
backup.

the Canadian astronaut team made its first visit to the
Johnson Space Center south of Houston, Texas, where
astronauts prepare for shuttle flights, and to the Kennedy Space Center in Florida, where shuttles are
launched. Garneau and Thirsk received NASA medical
examinations and were fitted for their flight suits.

Originally, Garneau was slated for a flight in late October. But the shuttle schedule for 1984 underwent many changes. His assignment was moved to an early October flight, which shortened his training period even more. In early August, Garneau and Thirsk reported to Houston for their final flight training.

When the shuttle program began in 1981, astronauts were already divided into pilot astronauts and mission specialists. Each shuttle flight is commanded by a pilot astronaut. Another pilot also flies on every mission. The pilots have a background flying jets and often have worked as test pilots, just like the overwhelming majority of astronauts who flew aboard the Mercury, Gemini and Apollo missions before the shuttle era. Mission specialists are assigned to operate scientific experiments or carry out other work on board the shuttle during flights. While a few scientist-astronauts flew before the shuttle, they still had to be trained as pilots. This requirement was dropped for mission specialists.

Garneau was to be one of the first of a new breed of astronaut—the payload specialist, a member of a space shuttle crew who is concerned with a specific payload. The astronauts at NASA regarded payload specialists with mixed feelings. Many veteran astronauts waited up to two decades for their ride into space on board the shuttle, while in some cases payload specialists were going into space ahead of them after just a few weeks of preparation. The first two payload specialists flew aboard a mission on the shuttle *Columbia* in late 1983, but they had prepared for months to operate sophisticated experiments on board the European

Space Agency's Spacelab module. One of them, German ESA astronaut Ulf Merbold, was the first non-American astronaut to fly on board the shuttle. Garneau would be the second. Another kind of payload specialist flew on board the flight immediately preceding Garneau's. Charles Walker, an employee of McDonnell Douglas Aerospace, became the first private sector astronaut when he flew aboard the shuttle *Discovery* to operate an experiment designed to produce new pharmaceutical products in weightless conditions. More payload specialists from varying backgrounds were contemplated for shuttle flights following Garneau's mission.

When Garneau and Thirsk began their training in Houston, they found that the offices for payload specialists were in a building well separated from the building where the other astronauts worked. The two Canadians were finally given space in the main astronaut office area as their training progressed so that they could be closer to their crewmates. Garneau had been assigned to a mission on board the shuttle *Challenger* with six other astronauts. Previous missions had carried crews of six; this crew of seven would be the biggest group launched up to that time.

Garneau describes his first look at the interior of the shuttle: "I said, 'my goodness, seven of us are going to be living in here for eight days?' I hadn't met these people at this time and I was thinking of them as strangers. I thought, this is going to be tough.' But during the two months, as I got to know them and became more familiar with the compartment, I felt more and more comfortable."

Garneau had crossed the Atlantic in cramped sail-boats. He would put that experience to work when he boarded the shuttle. The living areas include a flight deck and mid-deck, each the size of a small bedroom. Garneau later compared life onboard the shuttle to working in a submarine.

In the convoluted NASA system of the time, the flight was designated STS-41-G. The crew included two veterans, commander Robert Crippen and mission specialist Sally Ride, who gained fame on her first mission as America's first woman in space. Crippen joined the astronaut corps in 1969 and was the pilot of the first shuttle flight in 1981. Since then, he had commanded the seventh shuttle mission in 1983 and the 11th shuttle flight in April 1984, a mission that featured the dramatic capture and repair of the mal-functioning *Solar Max* satellite by the Canadarm and spacewalking astronauts. Also flying on 41-G were pilot Jon McBride and mission specialists Kathryn Sullivan and David Leestma. A few days after Garneau was assigned, a second payload specialist, Paul Scully-Power, was added. Sullivan had a Canadian connection because she earned her Ph.D. in geology at Dalhousie University in Halifax. Adding to crew unity was the fact that all crew members except Ride had a link to the sea—Crippen, McBride and Leestma were U.S. navy officers, Garneau was in the Canadian navy, Scully-Power was an oceanographer, and Sullivan also had a background in oceanography.

During those weeks, Garneau and Thirsk prepared for the mission by going for flights on a specially outfitted aircraft, known as the "vomit comet," that gives its passengers brief experiences of weightlessness. They also underwent simulations with other members of the crew as they rehearsed their work for the mission. When the shuttle was put up on the pad, the crew got on board to rehearse launch day, including a dash for the emergency baskets that would carry them from the launch tower to the ground in case they had to escape from the shuttle just before launch.

A group headed by Bruce Aikenhead (who worked on the Avro Arrow, helped train NASA's original seven astronauts and worked on Gerald Bull's HARP cannon, Canadian satellites and the Canadarm) was hard at work preparing the 10 experiments that Garneau would perform in space. Most of the scientific equipment had to fit inside a single locker in the space shuttle mid-deck, and everything had to be tested and approved by NASA safety experts before it could fly. "We literally set the record for packing the most stuff in a locker," Aikenhead said.

Finally, three days before the scheduled launch, the crew flew to Cape Canaveral to relax and make preparations. At a final barbecue, one day before the launch at a beach house near the launch pad, Garneau's wife and parents bid him farewell, along with other families who were visiting their loved ones.

Shortly before 3:00 AM on Friday, October 5, Garneau and his crewmates woke up in their quarters in the

Operations and Checkout building at the Cape, where every Gemini, Apollo and shuttle crew has spent their last night before launch. After showering, shaving and dressing, Garneau joined his crewmates for breakfast. During the early years of the shuttle program, crews did not wear pressure suits at launch, so they donned their blue flight pants and jackets and rode the elevator to the main floor of the building. They emerged to board the Astrovan (NASA's astronaut transport vehicle) in front of a crowd of media and NASA workers. The specially outfitted van took the seven astronauts 14 kilometres to the base of Pad 39A, the same pad used by most of the Apollo flights and all shuttle flights to that time. After riding the elevator 64 metres above the pad, the crew began boarding, starting with Commander Crippen and the three other astronauts who would be seated in the flight deck. Just after 5:00 AM, Garneau put on his helmet and crawled through *Challenger*'s hatch. He took his seat on the right in the mid-deck. Garneau was soon joined by Leestma and Scully-Power before the closeout crew closed the hatch and left the launch pad. For two hours, Garneau waited, his meditations interrupted only by occasional communications checks and some banter with other members of the crew because there are no controls in the mid-deck. The sequence went unusually well, and STS 41-G's countdown continued smoothly to the scheduled launch time—unlike many flights that suffer delays caused by technical glitches or weather problems.

The shuttle *Challenger* lifts off from Cape Canaveral with Canadian astronaut Marc Garneau on board, October 5, 1984.

A few seconds before the countdown clock hit zero at 7:03 AM, the three main engines at the base of *Challenger* lit up and gained power until the moment of launch, when the two white solid rockets flanking the vehicle's rust-coloured external tank roared to incandescent life. *Challenger* took to the skies for its sixth flight and the 13th launch of the shuttle program. Liftoff took place just at sunrise, and when *Challenger* pierced a bank of clouds that hung above the launch pad, the vehicle emerged into the first light of day, a scene that was witnessed by crowds of NASA workers, tourists and a Canadian television audience.

During the two minutes that the solid rockets remained attached to the vehicle, the crew experienced a ride that one astronaut compared to being shaken by the shoulders. Once the rockets separated from the shuttle and *Challenger* continued on its three main engines, the ride became much smoother. The astronauts were pressed into their couches as they accelerated toward orbit. Less than nine minutes after launch, *Challenger* and its crew were orbiting the Earth, 356 kilometres high, travelling at more than 28,000 kilometres per hour. The instant the main engines stopped firing, the seven astronauts entered weightlessness. This time, the sensation did not end after 30 seconds, as it had on the "vomit comet." Like the others, Garneau unstrapped himself from his seat and floated to the nearest window for an unforgettable first view of Earth from space. "The thing that struck me was how crystal clear the colours were," he later said.

Fewer than 200 people had gone into space before Garneau. Although he was only the second non-American to fly aboard the shuttle, the active guest cosmonaut program being pursued by the Soviet Union meant that Canada was the 14th country to have an astronaut in space. When *Challenger* departed Cape Canaveral that morning, it headed northeast toward an orbit tilted 57 degrees to the equator. It would be the second shuttle and only the fifth American spacecraft with astronauts on board to fly over Canada.

Once *Challenger*'s payload bay doors opened, the crew's main task on the first day was to release the *Earth Radiation Budget Satellite* (*ERBS*), which was designed to measure the solar radiation absorbed by the Earth's atmosphere and reflected back into space. Sally Ride used the Canadarm to grab the satellite and hold it above the payload bay while the satellite's solar panels extended. But the wings didn't open as planned, so Ride, working with mission control, shook the satellite using the Canadarm. When that didn't cause the panels to pop open, she pointed the satellite with the Canadarm so that the hinges holding the solar panels could warm up in the sunlight. That did the trick, and *ERBS* was released. During this time, Garneau and Scully-Power were ordered to stay out of the way in the shuttle mid-deck. Ride then took videotape of the satellite to test the Canadian-made Space Vision System (SVS). The underside of *ERBS* had been fitted with black target dots for the vision system's computers to follow. Garneau also filmed other targets set up in the shuttle cargo bay.

The satellite's balky solar panels were just the first of a series of problems the crew faced. Before the first day ended, the motors moving a high-power antenna used to transmit data also failed. The next day the crew executed a fix that stabilized the antenna but limited the amount of data that could be beamed back to Earth. The problem was even worse for one day when a relay satellite was temporarily lost. An important part of the crew's work involved running a large radar array in the payload bay that obtained images of the ground. When the shuttle lowered its orbit on the second day to 224 kilometres, the radar antenna wouldn't close properly, so Ride held it down with the Canadarm. The radar did obtain data from a pared-down list of sites, including fields in Saskatchewan, where local school children took soil samples to help calibrate the radar measurements. Another source of trouble during the flight was a balky cooling system. The fix involved allowing the temperature inside the cabin to rise to 32° C for a day.

While their crewmates dealt with these problems, Garneau and Scully-Power, an American citizen who was born and raised in Australia, were under orders to stay out of the way. But the two kept busy helping each other with their own work. Scully-Power's main task on the shuttle was to look for spiral eddies in ocean currents. When Scully-Power was photographing or otherwise recording these currents, Garneau, and sometimes one of the other astronauts, scouted for new eddies from a forward-looking window.

Garneau, for his part, was busy with 10 CANEX experiments, including the test of the Space Vision System. Another project, the Advanced Composite Materials Experiment, determined how a set of light-weight, non-metallic materials held up when exposed to the extreme conditions in space. These materials were attached to the Canadarm, and Garneau photographed them during the flight. In the OGLOW experiment, Garneau also took photos out the windows, this time with a camera fitted with a set of special filters designed to show the glow that has been observed on the shuttle's tail. Garneau's photos showed that the glow dissipates when the shuttle fires its thrusters. The camera was also used to photograph the upper atmosphere and auroral displays visible from space. In another experiment, Garneau measured solar radiation in the upper atmosphere using a photometer.

The remaining six experiments in CANEX supported Canadian research, most of it spearheaded by Ken Money and Douglas Watt, into space motion sickness, officially known as space adaptation syndrome. Garneau was to assess his health state, but he did not come down with space sickness. Another set of tests, which required assistance from Scully-Power, involved testing the vestibular-ocular reflex, which allows people to keep their eyes fixed on an object even when they are shaking their heads. The two payload specialists took turns shaking their heads to see how the reflex operates in space, while the other astronaut recorded the results. In one test, the subject wore a flashlight on the side of his head and covered his eyes. The experiment found that the reflex operates the same in space as on Earth, which means it is not related to space sickness.

Another major goal of the mission was rescheduled from the fifth day to the seventh because of the problems with the antennas and the cooling system. Sullivan and Leestma donned their space suits and went into space in the payload bay to test a refuelling system for use in future space operations. Along with

Ride, Sullivan made history on this mission because STS 41-G was the first spacecraft to carry two women. When Sullivan began her extravehicular activity, she became the first American woman to walk in space. A Russian cosmonaut had become the first woman to walk in space just a few weeks earlier.

Garneau was not involved in the spacewalk, but he and Scully-Power helped the crew with maintenance tasks, such as preparing meals and changing the cartridges that clean the shuttle's air of carbon dioxide exhaled by the astronauts. They also appeared in photos and motion pictures taken by the other crewmembers, including film taken with the Canadian-built and designed IMAX large-screen motion picture system.

Garneau, who fit in with the crew because of his similar background to the American astronauts, also worked hard to make sure that his work did not interfere with other mission goals. For Garneau, that included not talking to the ground except when necessary. Garneau's non-communication came to bother the large group of Canadian journalists that had gathered at Mission Control in Houston to cover the flight. One reporter was so upset that he branded Garneau "the Right Stiff" because of his relative silence. "He was very much feeling his place as a payload specialist as he had learned it from NASA—don't be in the way, do your experiments, and he might or might not get some help from other people," said Aikenhead, who was monitoring the flight from the control centre. "So Marc tried, as a good navy man, to stay out of the way. We actually had to nudge the system a bit to get a daily voice report from him."

Finally, the crew held a press conference on the fifth day of the flight for a group of reporters, including Garneau's brother Phillippe, who was covering the mission for a radio network in Québec. Garneau described how he felt the noise and vibration of launch and admitted that he was "a little bit afraid." In orbit, his strongest impressions were of looking back at his home country. "Every time we go over Canada, I'm generally stuck at the window, having a look out there." Like many astronauts, the word he used to describe the view was "fantastic."

During the flight, residents of Kingston, Ontario, flashed their lights as *Challenger* made a predawn pass over the city, and Garneau radioed his appreciation. And one morning, Garneau replied to Houston's wakeup call with a humorous acknowledgement in French that baffled the unilingual capsule communicator—a French version of the recording on automated phone systems when the line is busy but someone will answer soon. The crew spoke with President Ronald Reagan, but Prime Minister Brian Mulroney, who took office just days before the launch, did not take the opportunity to speak to the astronauts.

Finally, on October 13, after eight days in space and 132 orbits of the Earth, *Challenger* and its crew returned to Earth at the landing strip near the launch site at the Kennedy Space Center. Preparations began a few hours before landing. The crew put away their equipment, unstowed their seats and donned their flight suits and helmets. Then the engine fired to slow the shuttle enough that it began its descent from orbit to a point where the atmosphere would further slow the space-craft down. While the shuttle streaked across central Canada and the U.S. on the way to Florida, gravity pressed the astronauts to their seats. They wore pres-sure cuffs on their legs to prevent blood from pooling there. STS 41-G was only the second shuttle to land at the Florida launch centre, which pleased Commander Crippen. His two previous missions had been diverted from Florida to Edwards Air Force Base in California because of bad weather at the Cape.

After the landing at 12:26 PM, the astronauts remained inside *Challenger* for a half hour. When Garneau stood up for the first time in gravity since launch, he nearly fainted. But he found his feet and left the shuttle with his crewmates for a round of med-ical examinations and follow-up tests on the space adaptation experiments he'd performed in space. That day, the crew returned to Houston. After the family reunions and post-flight debriefing sessions, Garneau and his wife took a holiday in the Bahamas before he began a task that was almost as daunting as going into space: touring Canada to share his experiences.

The tour began in Ottawa with a dizzying round of media interviews. Garneau and Thirsk made public appearances across Canada, where they showed films of their training and Garneau's flight. In each presen-tation, Garneau used the word "fantastic" several times to describe his experiences. Garneau served as

Grand Marshall of the Grey Cup parade in Edmonton, and other appearances included meeting politicians and receiving awards. As the tour wound up in Toronto in early December, Garneau marked one year as an astronaut with an appearance with his shuttle commander, Bob Crippen, who praised Garneau for doing a "superb job." He added, "I would be pleased to fly with Marc or any other Canadian astronaut."

CHAPTER TWELVE

Roberta Bondar and the
Return to Flight

IN EARLY 1986, THE U.S. HUMAN SPACE PROGRAM WAS HEADED for its most ambitious year since the end of the Apollo Moon landings. NASA planned 14 shuttle flights for 1986, including a mission that would feature observations of Halley's Comet and launches of the *Hubble Space Telescope*, the *Galileo* probe to explore Jupiter, and the *Ulysses* spacecraft to orbit around the poles of the Sun. The latter two spacecraft were to use powerful and dangerous liquid-fuelled Centaur rocket stages to propel them from the shuttle in low Earth orbit into their new paths. Another flight carrying a secret payload for the U.S. military was planned for launch into a polar orbit from a brand-new launch pad in California. Other shuttle flights were to carry other military satellites as well as commercial payloads such as communications satellites.

NASA launched nine shuttle flights from Cape Canaveral in 1985. Most of them carried communications or military satellites or focused on scientific research. Many payload specialists flew that year, including two engineers employed by American corporations, a United States senator, foreigners from Europe and Mexico and a Saudi Arabian prince. The first shuttle flight of 1986 carried a member of the U.S. House of Representatives and a privately employed engineer on board *Columbia*. The second mission was carrying another corporate passenger and, famously, the first ordinary U.S. citizen to fly into space: schoolteacher Christa McAuliffe from New Hampshire.

The flights of Canadian astronauts originally scheduled for 1985 and 1986 had been shifted back to 1987. But as 1986 opened, Steve MacLean was in training, along with his backup, Bjarni Tryggvason, for a 1987 shuttle flight in which he would test the Space Vision System. The three medically trained astronauts, Roberta Bondar, Ken Money and Bob Thirsk, were hoping to be named for another shuttle mission that would further advance Canadian expertise in space motion sickness.

Late in the morning of January 28, 1986, *Challenger* was launched on its 10th mission, the 25th flight of the shuttle program. It carried a communications satellite, a science payload and crew of seven, including McAuliffe. Beyond the school children who were looking forward to McAuliffe's lessons from space, the launch attracted little public notice beforehand.

That day, Marc Garneau was attending a meeting at the Johnson Space Center in Houston. In Ottawa, Steve MacLean walked from his office to his nearby home to watch the launch. Garneau's meeting paused to watch the launch and was about to return to business when the two solid rockets boosting *Challenger* split off prematurely and formed a peculiar Y-shaped cloud. Everyone in the room knew instantly that something was seriously wrong, but no one acknowledged it until some began to cry. MacLean got home too late to see the launch live, but saw replays as soon as he switched on his television. "I can remember thinking what my parents would feel if it had been me," he recalled. Money walked into the astronaut office in Ottawa a few minutes later and was told that the shuttle had exploded. He didn't believe it until he turned on a radio. When Bondar joined him a few minutes later, she instantly believed Money's news, because he wouldn't joke about such a thing. That night, MacLean went to see his parents, who encouraged him to keep up with his plans. But the *Challenger* explosion changed everything for the shuttle program and Canada's astronauts.

All shuttle flights, including MacLean's flight and the other planned Canadian flight, were cancelled until the cause of the *Challenger* disaster could be found and the necessary changes made. Although the Canadian astronauts got the good news a few weeks later that Canada intended to join in the American-led space station program, which would mean more flights in the future, it soon became clear that the hiatus in the shuttle program would not be a short one. In June that year, a presidential commission pinpointed the cause of the explosion—the failure of an O-ring in a solid rocket motor. But the cause was also related to a culture within NASA, where problems such as O-ring leaks were allowed to fester until they cost lives. The shuttle program required major changes in management and in safety measures. It was nearly three years after the *Challenger* disaster before the shuttle *Discovery* took to the skies in the fall of 1988, marking the resumption of shuttle flights.

Among the many changes to the shuttle program was an end to commercial payloads. Commercial launches were shifted to traditional rockets. The U.S. Department of Defence withdrew from the shuttle program after the completion of a handful of missions that could only fly on the shuttle. Launches of space probes and research satellites also were shifted away to unmanned rockets. Once the backlog from 1986 was cleared away, the shuttle was to concentrate on scientific research and flights to the space station once the station was launched in the 1990s. And the payload specialist program was to be scaled back—a move that affected Canada's six astronauts and lengthened their waits to get into space.

While they waited to get new flight assignments, the Canadian astronauts concentrated on their research or on education rather than training for flight. Bondar and Thirsk, the two physicians, began doing part-time medical work in hospitals to keep up their skills while they continued their research. Money went back to work on space motion sickness. MacLean, Tryggvason and Garneau continued with preparations for MacLean's flight at first, but the trio then shifted to research work on the Space Vision System, especially MacLean, who worked on upgrading the system. Thirsk, Tryggvason and MacLean also began to have children and raise families. Garneau faced a formal separation from his wife Jacqueline, followed by her tragic death by her own hand a few months later, making him the single parent of their twins.

Early in 1989, with the shuttles back in flight, NASA announced a new schedule that included a research mission, planned for 1990, that would focus on the effects of weightlessness on astronauts. Canada was invited to fly an astronaut on the mission. Bondar and Money began preparing for the flight. MacLean was told that his flight was postponed to 1992. During the year that followed, Bondar and Money began preparing for one of them to fly in a friendly but sometimes uneasy competition. In January 1990, 200 scientists from 13 countries involved in the mission voted for Bondar as the Canadian participant in the first International Microgravity Mission (IML-1), with Money as her backup.

In contrast to Garneau's abbreviated training period, Bondar, Money and the crew of the IML-1 spent years preparing for the flight, which was planned before the *Challenger* disaster. The launch was postponed 19 times as the shuttle and the crew were changed, including the tragic loss of popular astronaut Manley "Sonny" Carter in the crash of a commuter aircraft in April 1991. Finally, the crew's training was completed in January 1992. *Discovery* stood ready on Pad 39A at Cape Canaveral.

The European-built Spacelab took up much of *Discovery*'s payload bay. The module was a pressurized space fitted with equipment where the astronauts

Canadian astronaut Roberta Bondar

~•∞)(∞•~

could conduct scientific work. IML-1 carried a series of experiments to examine the effects of space travel on humans and of the space environment on biological materials, and another set of experiments to create crystals in weightlessness, including products for use in computers and proteins that could be used to treat health problems.

Six Canadian space physiology experiments were designed to look at the mysterious mechanisms behind space sickness (space adaptation syndrome).

The sickness is thought to result from conflicting inputs from visual cues and organs in the inner ear that control the body's balance and sense motion and position. Both Money and Douglas Watt of McGill University developed experiments used on IML-1 and other flights, including Garneau's 1984 mission.

On the morning of January 22, the crew of IML-1, also known as STS-42, walked out of the Operations and Checkout building wearing the orange pressure suits that all shuttle crews since the *Challenger* explosion have worn at liftoff and re-entry. Before stepping aboard the Astrovan to go to the pad, Bondar waved her arms in exultation.

The journey to the launch pad marked a moment Bondar had anticipated—and worked toward—for the 46 years of her life. Growing up in Sault Ste. Marie, Bondar witnessed the early days of the space age. Like many other kids her age, she grew up following the early space flights and reading science fiction comics. She and her older sister Barbara built model rockets and pretended to be space explorers.

"When I was young, I wanted to be an astronaut so badly. Of course as a child, you want to be a lot of things," she recalled shortly before her launch. "I wanted to be a doctor, an astronaut and a scientist."

The young Bondar was an active athlete in school as well with the Girl Guides and on camping trips with her family. Unlike many people of the time, Bondar's parents, Edward and Mildred Bondar, did not discourage their daughters from pursuing whatever they wanted to do. They stood behind Roberta when teachers tried to turn her away from studying science. Once, she was denied the position of school patrol captain in favour of a boy, even though she had done best in the test for the job. While this incident and others caused her to fight discrimination against women, she also resists the idea that she should get ahead simply because she is a woman.

"I don't think a great deal about my sex or my nationality," she said when she was selected for the flight. "What motivates me is to go into space, and above all, it's representing my profession by being the first neurologist to go into space."

Bondar decided to pursue science as a career after being inspired by a research project on the life cycles of caterpillars. She earned a bachelor's degree in zoology and agriculture at the University of Guelph, followed by a master's degree in experimental pathology from the University of Western Ontario in 1971, a doctorate in neurobiology from the University of Toronto three years later, and finally her medical degree from McMaster University in 1977. She was accredited as a specialist in neurology in 1981.

In 1983, when she heard on the radio that the Canadian Astronaut Program was looking for recruits, she was an assistant professor of neurology and the director of the multiple sclerosis clinic at McMaster. While her career path may appear to have steered her away from space, in her research she included the organs in the inner ear related to space sickness.

She also pursued active hobbies, most notably flying and hot air ballooning, along with canoeing, biking, target shooting, fishing, cross-country skiing and the favourite astronaut sport, squash. But Bondar never had a chance to begin a family. "I thought I could spend two or three years in the [space] program and still have time to have children," said Bondar, who was 38 when she became an astronaut and 46 when she finally got her chance to fly into space.

When Bondar eased herself into *Discovery* for launch, backup Ken Money was waiting at the payload operations control centre for IML-1 at Huntsville, Alabama. "The scientists involved in this decision did vote and did elect Roberta—she won on that basis fair and square," he said when Bondar won the seat on IML-1. For much of the mission, Money would be Bondar's link with planet Earth.

While the two astronauts remained friends, the competition for the flight was long and occasionally testy. When Money was selected as backup, he announced that he would leave the astronaut program after the flight. On launch day, Money was 57, and he had conducted research in high altitude and space medicine for 30 years at the Defence and Civil Institute of Environmental Medicine in Toronto, where he would return after his stint as an astronaut. Even before he became an astronaut, Money was a familiar figure at NASA because of his work on space motion sickness and orientation problems. He also worked part-time at the University of Toronto and put out more than 80 scientific publications.

Bondar's flight received almost as much coverage as Garneau's mission seven years earlier. The fact that Bondar was Canada's first woman to fly into space and the long hiatus between her flight and Garneau's mission meant that Canadians would be tuning in to follow her progress.

Discovery lifted off an hour later than scheduled because of a balky fuel cell and an erroneous warning of weather problems. Just eight minutes after liftoff, the seven astronauts of IML-1 were in an orbit that, like Garneau's flight, made daily passes over Canada. But in many other ways, this mission was different. This time, the shuttle carried the Spacelab module, which vastly increased the space available to the crew of seven. And the crew would operate around the clock with two shifts of 12 hours each. Bondar was on the "red" team of astronauts and controllers, who worked the day shift. Once the shuttle was in orbit and the crew had stowed the pressure suits and seats, Bondar and her mates on the red shift were the first inside the lab, getting the experiments started.

For astronauts working inside the lab, a specialized control centre for scientific payloads established in Huntsville was the point of contact rather than the usual control centre in Houston. Money was Bondar's contact person. "We had to send her to lunch," Money said during the first day of operations. "She was working through her lunch break."

Much of Bondar's time was spent on space sickness experiments that built on the work that Garneau had done, such as tests of the relationship between visual and other cues in space adaptation syndrome. The lab was big enough to contain a sled designed by Watt and his team and McGill and a rotating chair similar to that used by Money at his institute in Toronto to induce motion sickness. Space sickness has been puzzling to scientists and astronauts because people who are susceptible to motion sickness on Earth may not suffer from space sickness and vice versa. The malady usually strikes in the first few days in space, which strongly affects shuttle missions because they usually only last a few days. Space sickness is such a touchy subject amongst the astronauts that NASA refuses to say whether any astronaut gets space sickness during a flight. Research into space sickness is intended to benefit more people than just the astronauts. Watt works with other researchers who deal with dizziness and balance problems on Earth. Using the sled and the rotating chair on the shuttle allowed comparison of results between users on Earth and in space.

The crew also looked at other medical problems that affect astronauts. For example, when astronauts go into space, they gain up to five centimetres in height because the lack of gravity allows the spaces between the vertebrae in the back to spread apart. But this condition, which ends upon return to Earth, also causes back pain for astronauts. On IML-1 and other flights, astronauts took measurements of their backs and recorded their pain levels for a team at the University of British Columbia in Vancouver. Bondar reported that she temporarily grew four centimetres in space, and she was surprised to find that the changes caused by zero gravity also allowed her to put away her glasses, at least until she got back to Earth.

University of Calgary scientists sponsored a test of how much energy astronauts expend in space. Canadian astronaut Bob Thirsk designed an experiment to test an anti-gravity suit that counteracts the effects of long space flights on the cardiovascular system by applying pressure to the legs and lower abdomen.

Bondar and her colleagues also conducted experiments on crystal growth and on the development of various forms of life, including frogs' eggs, bacteria and yeast, to see how they grow in weightlessness. The mission's 42 experiments led Bondar to compare her flight to one big open-book exam.

But the flight wasn't all work. Bondar had time to go to the windows and look at the Earth, Canada and her beloved hometown, Sault Ste. Marie. "As I look down, across, and above from the flight-deck window, the shining planet curves from left to right," she wrote in her memoir, *Touching the Earth*. "I have never in my life seen anything as big as this. The sheet of oceans and land masses moves like a rolling ball edged in black, but from this altitude I can never see the whole of the ball at one time."

She noted the many colours on the Earth and the various shades of blue in the oceans. The thin atmosphere protecting life on Earth has a three-dimensional quality that can't be reproduced in photographs, and above is a black she called profound. Her book is full of poetic descriptions of the Earth from space and how it compares to the view from the ground.

"It's been a fantastic experience. I can't wait to get back to tell people about it," Bondar told Prime Minister Brian Mulroney when they chatted near the end of the flight. "We certainly are all of one planet."

Bondar and her crewmates also spoke to President George H.W. Bush, and since the flight coincided with Super Bowl XXVI, the crew appeared during the pregame show. Bondar clutched a coin as her crewmates spun her around. "Looks like heads," one television commentator said when Bondar's double flip ended. "In zero-G, the coin never comes down," pilot Stephen Oswald said. "So we'll have to defer the coin-toss to the official pre-game ceremony."

While the mission had some minor equipment problems, the crew conserved enough power extend the flight from the scheduled seven days in space to eight. On January 30, the astronauts shut down the laboratory, donned their pressure suits and rode *Discovery* to a successful landing at Edwards Air Force Base in California. Back on the ground, the crew had to wait to shower because they were subjected to a set of medical tests to follow up on their measurements taken in orbit.

Bondar came home a hero, particularly in her hometown and for girls who look up to her as a role model. But a few months after her flight, Bondar left the astronaut program, in part because there appeared to be little opportunity for her to get another space flight or to continue her research program. Despite facing problems finding support for her work from Canadian universities, Bondar continued her medical research. She still makes public appearances where she shares her experiences in space, something she also did in her memoirs and a children's book she wrote with her sister, *On the Shuttle: Eight Days in Space*. Bondar has also transformed her passion for photography into three acclaimed books of landscape photographs.

CHAPTER THIRTEEN

Steve MacLean Tests the Space Vision System

WHILE ROBERTA BONDAR WAS GETTING USED TO LIFE BACK ON Earth, Steve MacLean and backup Bjarni Tryggvason were preparing for MacLean's long-postponed flight to test the Space Vision System. When MacLean lifted off with five other astronauts on board *Columbia* on October 22, 1992, his flight was treated as routine by the Canadian media, who were not excited by the second Canadian astronaut flight of a busy year—a year that included the first launch of *Challenger*'s replacement shuttle, *Endeavour*, and its use in a dramatic satellite salvage operation. MacLean's flight also coincided with the Toronto Blue Jays playing in their first World Series and Canadians voting on the Charlottetown constitutional accord.

MacLean, a strong Blue Jays fan, did not mind. He was relieved to see an end to the postponements that had stood in the way of his flight into space, and he happily collected on wagers when the Jays defeated the Atlanta Braves to win their first championship. After nearly nine years in the astronaut program, the 37-year-old native of Ottawa who earned his bachelor of science and his doctorate in laser physics from York University in Toronto was glad just to fly. The wiry

MacLean also took active part in gymnastics at York and made the national team, just missing a spot at the Olympics. When he joined the astronaut program in 1983, he was continuing his laser physics research at Stanford University in California.

In the astronaut program, MacLean went to work with Lloyd Pinkney, the NRC engineer from Alberta who developed the Space Vision System (SVS) for the Canadarm. The SVS links computers with television cameras shooting targets on satellites or other vehicles to provide real-time computer images of where the Canadarm is in relation to the object it needs to grasp. SVS is particularly useful when the object cannot be seen directly by the astronaut operating the Canadarm because something is in the way or because of the harsh lighting conditions in space.

MacLean's father worked at the National Research Council. The young MacLean first met Pinkney and saw him working on a predecessor to the SVS long before the astronaut program started. When his 1987 flight was postponed after the loss of *Challenger*, MacLean worked with Pinkney and other NRC scientists on improving the SVS.

In orbit, the STS-52 mission launched a LAGEOS satellite that reflects laser beams for use in tracking continental drift and changes in Earth's gravitational fields. After the launch, *Columbia*'s crew set to work on experiments similar to those carried out on Bondar's mission but more modest because the STS-52 didn't carry a Spacelab module. MacLean worked on

The crew of the STS-52 shuttle flight. Top: L to R: Pilot Michael A. Baker, Commander James D. Wetherbee, Canadian astronaut Steven G. MacLean; bottom, L to R: astronauts Charles L. Veach, Tamara E. Jernigan, William M. Shepherd.

experiments known as CANEX-2. Many of them built on the experiments carried out by Garneau on his flight, including a repeat of Garneau's attempt to determine the cause of the glow around the shuttle's tail in orbit. Another experiment from Queen's University in Kingston involved melting metal in a small furnace to help create better alloys on Earth. MacLean pointed a photo-spectrometer outside a window that did not have protective coatings and got sunburned for his efforts.

Materials attached to the Canadarm were exposed to the radiation and atomic oxygen 300 kilometres above the Earth (where the shuttle flies). A similar experiment was carried out on Garneau's flight. NRC investigators were surprised by the damage sustained by some of the materials, including plastics and paints. One of the researchers was MacLean's father, NRC chemist Paul McLean, whose last name is spelled differently from the rest of his family because of a clerical error on his birth certificate. The elder McLean developed a material tested on his son's flight for possible use in space station components.

A week into the flight, MacLean was working so hard that his commander lightened his load by delegating some of it to other crewmembers. "Steve's doing a great job," commander James Wetherbee said. "He wants to do it all." During a television interview on the CBC, MacLean said, "I hardly got anything to eat because we were so busy."

The most important Canadian experiment, the Space Vision System test, was performed near the end of the flight. Astronaut Charles Lacy Veach used the Canadarm to unberth and berth the Canadian Target Assembly (CTA), a large metal object simulating a satellite covered with targets for the SVS. MacLean used the SVS to help Veach berth the CTA, while the SVS recorded their work. MacLean said that at times, he pretended the CTA was a truss for a space station. Finally, Veach released the CTA while the SVS continued to follow this latest Canadian satellite until it drifted away. The SVS worked effectively, but there were problems when sunlight suddenly changed brightness levels in the television camera.

Canadian astronaut Steve MacLean floats inside the shuttle Columbia during the flight of STS-52.

~⊶✿⊷~

After nearly 10 days in orbit, STS-52 ended successfully on the runway at Cape Canaveral. The conclusion
of his flight marked the conclusion of a successful return
to flight for Canada's astronauts. But changes were
sweeping through the Canadian Astronaut Program,
including its transfer from the NRC to the newly formed
Canadian Space Agency (CSA).

Bondar and MacLean flew as payload specialists, but
few payload specialists were flying any more and their
roles had changed because training was stepped up. The
days were over when a payload specialist was little more
than a glorified passenger, as was the case in the months
between Garneau's flight and the loss of *Challenger*.

In 1990, NASA invited the CSA to begin training its astronauts as mission specialists, who operate shuttle systems including the Canadarm and carry out space-walks. Mission specialists, who until that time had only been Americans, undergo a year of intense training in Houston before being qualified to fly. The CSA decided to send two Canadian astronauts for mission specialist training in 1992. It also decided to select new astronauts to enlarge the Canadian astronaut team in preparation for the space station program.

A few days before Bondar's flight in January 1992, the CSA advertised in newspapers for astronauts with engineering and life sciences backgrounds. This time, 5330 people applied. Like the 1983 selection process, most of the applicants were weeded out in the early reviews. The final 50 went to Ottawa for a round of tests including questioning by a seven-member committee, including Marc Garneau and Bob Thirsk.

Finally, in June, the CSA announced that four people had been selected. They were Julie Payette, a talented computer engineer working at Bell Northern Research; Chris Hadfield, a Canadian Forces jet pilot who had worked with top American test pilots; Dafydd "Dave" Williams, an accomplished emergency room physician in Toronto; and Robert Stewart, a geophysicist from the University of Calgary. Stewart surprised the committee two weeks after the announcement when he decided not to join the astronaut program and instead opted to remain in Calgary. He said his decision was influenced by continuing uncertainty over whether the recruits would get to fly and his commitment to

Canadian astronauts in 1998: Front L to R: Chris Hadfield, Bjarni Tryggvason, Steve MacLean. Back L to R: Robert Thirsk, Marc Garneau, Dave Williams, Julie Payette.

his existing career. When Stewart announced his decision, the CSA called the top person who had not been selected, Canadian Forces computer engineer Michael McKay. In 1995, McKay resigned from the astronaut corps for medical reasons, but he continued to work for the CSA until he returned to the military in 1997.

Once the new astronauts were selected, CSA announced that Garneau and Hadfield would move to Houston to begin mission specialist training. Hadfield's selection proved controversial among the original six astronauts, but over the next few years, the remaining Canadians also underwent training as mission specialists, making them full-fledged astronauts in the eyes of NASA, ready for flight on what would become the International Space Station.

CHAPTER FOURTEEN

Chris Hadfield and Space Stations

WHEN CHRIS HADFIELD BOARDED THE SHUTTLE ATLANTIS ON its launch pad on November 12, 1995, he was blazing several new trails for the Canadian astronaut team. Hadfield was the first member of the second intake of Canadian astronauts to fly, and he was the first Canadian to fly as a mission specialist. Unlike Canadians who had flown before him, Hadfield was preparing for launch on the flight deck, which meant he would be helping the commander and pilot once *Atlantis* lifted off. He would also have a better view of the ascent than astronauts who sit in the mid-deck. Along with Garneau, Hadfield spent the previous three years in astronaut training at the Johnson Space Center in Houston. Part of that training involved serving as capsule communicator (CAPCOM), talking to other astronauts when they were in space, or preparing crews for launch at Cape Canaveral as a member of the astronaut support personnel—the "Cape Crusaders." But now it was Hadfield's turn, and he had a central role in one of the most ambitious shuttle flights ever.

When NASA first designed the shuttle, the agency hoped to see the spacecraft live up to its name by moving astronauts and equipment between the Earth and space stations. But the U.S. government gave the go-ahead to the shuttle in 1972 without a space station, so most flights either carried experimental laboratories, such as Roberta Bondar's flight, or they dropped off satellites and on occasion repaired them. In 1984, U.S. President Ronald Reagan finally told NASA it could start building a space station. But by the time Hadfield prepared for launch nearly 12 years later, Reagan's vision had undergone major changes. Originally named Freedom in a jab at America's Cold War adversary, the Soviet Union, the station proved much more expensive and difficult to build than expected. The Soviet Union collapsed in 1991. In 1993, NASA joined with the Russian Space Agency that emerged in post-communist Russia to work together on a redesigned station—the International Space Station.

At the time, the Russians already had a space station in orbit, so the two countries decided to make use of the Russian Mir space station to prepare the shuttle and American astronauts for work on the ISS. As part of the Shuttle–Mir program, Hadfield and the other four astronauts on STS-74 were going to Mir.

Mir was just the latest in a long line of space stations launched by the Soviet Union. Both sides in the Cold War dreamed of space stations. The idea of building stations in Earth orbit goes back to the 19th century. Early 20th century spaceflight theorists such as the Russian Konstantin Tsiolkovsky and the German Hermann Oberth created designs for space stations. Wernher von Braun, the German rocket pioneer who took a leadership role in the U.S. space program after

World War II, became famous for his vision published
in magazine articles and shown on television of voy-
ages to the Moon and Mars— a gigantic wheel in
Earth orbit that would serve as a way station between
the Earth and other celestial bodies. The wheeled
space station also appeared in Stanley Kubrick's
famous 1968 science-fiction film, *2001: A Space Odyssey*.
At the time, the U.S. and the Soviet Union were rac-
ing to the Moon, but they were also preparing to build
the first stations in space.

So what are space stations good for? Originally, they
were seen as a place to watch activities on the Earth
or point a telescope at the heavens, but robot satellites
have taken over those functions. Stations are also
places where astronauts can carry out experiments
such as creating new materials in weightlessness or
learning about how the conditions of space affect the
human body. Ultimately, space stations have been
seen as way stations in low Earth orbit for travel to
the Moon and other celestial bodies. In *2001: A Space
Odyssey*, for example, a vehicle delivers passengers
from Earth to the space station, and then spacecraft
for the Moon and the planets depart from the station.

The first real space station was the Soviet Salyut 1
that hosted a crew of three cosmonauts for 24 days in
June 1971. The mission had a tragic end. When the
three cosmonauts died on the way home when their
ferry spacecraft sprung a leak in the vacuum of space.
The Soviets pressed on with a whole series of Salyut
stations, but they found that the road to space is long
and difficult. Some Salyuts exploded or spun out of

control before cosmonauts could visit. Some cosmonaut crews could not dock with the space stations and had to make emergency landings. One crew nearly drowned when their returning capsule broke through the ice of a frozen lake while a blizzard held rescuers at bay. Two crews survived accidents during launch.

Even once they made it inside the station, some cosmonauts found that they had trouble getting along. One mission was cut short because of strange and unexplained odours. Another cosmonaut had to come home early when he got sick. The Soviet space program learned that crews have a lot of work to do just to keep their stations running. Soviet cosmonauts succeeded in completing daring repairs on broken equipment, and they learned the art of orbital refueling.

The Americans had a space station of their own, Skylab. The station was built from a surplus third stage for a Saturn V rocket and was launched in May 1973, shortly after the final Apollo Moon landing. Even before Skylab got into orbit, a micrometeoroid shield unexpectedly deployed, taking a vital solar panel with it as it fell off the station during the launch phase. The first crew of Skylab had to install a new shield and free a stuck solar panel in a series of daring spacewalks to save the Skylab program. Two more crews flew to Skylab that year and set a record for the longest human space flight up to that time—81 days. The Soviets took a few years to break that record, but by the mid-1980s, they were far ahead of the U.S. in terms of space station experience.

Reagan was inspired to begin the U.S. space station program in part because of the growing Soviet lead in that field. NASA didn't waste any time looking for international partners to join the program, including Japan, the European Space Agency and Canada. In 1985, Canada agreed to join, and Europe and Japan signed on to build scientific research modules for the station. The Canadian government studied several options for the station and decided on an advanced version of the Canadarm known as the Mobile Servicing System. Negotiations with NASA were complicated, because some people in NASA and the U.S. Congress wanted to have Americans build all parts of the station, but the difficulties were overcome, and the two countries reached an agreement in 1986.

Within a few months, the loss of *Challenger* set NASA and the space station program on its heels. As time went on, problems began to mount up. Infighting gripped NASA as its centres in various parts of the country fought for control over the project, each of them backed by congressmen seeking new economic activity in their districts. The awkward management structure NASA set up for the station produced expensive paper studies as costs exploded well beyond the original $8 billion price tag.

In 1993, the newly elected president, Bill Clinton, ordered a major redesign to bring costs under control. Clinton acted just as Congress came within one vote of cancelling the program. Also, the collapse of the Soviet Union opened up the possibility of including Russia as a partner, something that became a reality when Russia agreed to come on board later that year. The space station was no longer known as Freedom; it became the International Space Station.

Canada's contribution to the station also faced some political problems, most of them related to Canada's large federal deficit of the time. When Prime Minister Brian Mulroney's Conservative government agreed to take part in the station in 1985, it did so as it was cutting other science programs as part of the battle against the deficit, a fact that angered many scientists. At least once, the government considered withdrawing from the project because of mounting costs. When the Liberals returned to office in 1993, with the deficit still a problem, the new finance minister, Paul Martin, decided to pull Canada from the station program. The announcement was printed in the 1994 budget document, but a last-minute warning to NASA resulted in a personal intervention by President Clinton with Prime Minister Jean Chrétien. The Liberal government decided that Canada would remain in the program.

Development of the Mobile Servicing System was given to a group of Canadian companies that had been involved with the Canadarm, starting with prime contractor Spar Aerospace and major subcontractors CAE Electronics Ltd., Canadian Astronautics Ltd. and SED Systems Inc. The contractors worked under a group at the NRC headed by Karl Doetsch, and the team moved to the Canadian Space Agency when it was established in 1989.

The Russian Mir space station in orbit

~∞)(∞~

The Mobile Servicing System consists of three parts. The first, an advanced version of the shuttle Canadarm, is capable of moving end over end like an inchworm. The arm, naturally known as Canadarm2, is usually attached to the second part of the system, the Mobile Base System. The third part, which is also carried by Canadarm2, is called Dextre, the Special Purpose Dexterous Manipulator (or "Canada Hand"), a device capable of fine movements and manipulations. The system can carry out service in areas on the exterior of the station that would otherwise have to be carried out by astronauts on hazardous spacewalks.

The station's redesigns in the 1980s and early 1990s meant changes for Canada's contribution to the station. U.S. costs soared as time went on, and so did the cost of the Mobile Servicing System, which rose to $1.4 billion from the original estimate of $800 million.

Reagan had hoped to have the station in orbit by 1994, but the first components would not be ready to fly until later in the decade. So NASA and its new Russian partner agreed to get to know one another better by flying American astronauts on the Russian Soyuz spacecraft and Russian cosmonauts on the U.S. shuttle. As well, astronauts would join Russian crews on board the Mir space station for extended stays in space. By 1994, Russian cosmonauts were routinely spending months aboard Mir, while the longest American stay in space remained the 81 days of the final Skylab crew. One Russian had spent 438 uninterrupted days in space, and two others spent a year in space.

The first component of Mir was launched early in 1986, just a few days after the *Challenger* disaster grounded NASA's shuttles. By then, Soviet cosmonauts had great experience with space stations. The first cosmonaut crew to visit Mir was able to fly from Mir to an inert Salyut 7 station, revive it and transfer equipment over to the new station. While the Mir base block was similar to its Salyut predecessors, it carried a special docking adapter that allowed up to five other modules to join it, while still leaving two docking ports open for Soyuz ferry craft and Progress supply vehicles. By 1994, three modules had been joined to the Mir base block. The economic problems that led to the collapse of the Soviet Union left two other large Mir modules sitting on the ground. The U.S. funding that the Russians gained from the Shuttle-Mir program led to the installation of the Spektr module in 1995 and the final module, Priroda, in 1996.

In 1995, the Shuttle–Mir program began in earnest with a shuttle rendezvous with Mir in February and the launch of the first American astronaut to live on board Mir—Norman Thagard—in March. At the end of June, *Atlantis* was dispatched from Cape Canaveral to pick up Thagard and his Russian crewmates on Mir and replace them with two cosmonauts. The flight marked the first docking between Mir and a shuttle, but Mir's modules had to be moved around before and after the flight—a difficult operation—to make the docking possible without *Atlantis* bumping into Mir's solar panels.

NASA and the Russians decided that the solution was to add a small docking module to Mir to allow regular visits from the shuttle. The new module would be delivered by the crew of STS-74, including Chris Hadfield. He would have the toughest job on the mission.

The fact that NASA was giving a rookie astronaut from Canada such responsibility surprised people who didn't know the 36-year-old Hadfield. But even though Hadfield has shown extraordinary drive and is a member of Mensa, the organization for people with high IQs, his mother Eleanor described him as a "happy child, an ordinary average bright kid."

Canadian astronaut Chris Hadfield

Born in Sarnia, Hadfield comes from a family where flying comes naturally. His father Roger and two brothers were pilots for Air Canada. Like his brothers, Chris Hadfield joined his father at a young age, flying in small aircraft that the elder Hadfield kept on the corn farm he operated near Milton, Ontario, where the family moved when Chris was seven. Everyone had to help out with the corn crop, and Chris learned to fix balky tractors and other equipment.

When Hadfield was 10 years old, he watched Neil Armstrong and Buzz Aldrin make the first walk on the Moon. The boy decided he wanted to be an astronaut. He joined the Air Cadets and earned his pilot's license when he turned 17. After finishing high school, he and a friend took a long trip around Europe. While travelling, Hadfield decided he still wanted to be an astronaut, even though the U.S. space program was in the long hiatus between the Apollo and shuttle missions and prospects for Canadians to fly in space were unknown at best. Hadfield returned home and joined the Canadian Forces. He studied engineering at Royal Roads Military College near Victoria and the Royal Military College at Kingston.

Hadfield turned to a career in the air force. Through his skill and hard work, he won a job flying a CF-18 interceptor aircraft. He was stationed with No. 450 Squadron at CFB Bagotville guarding the Atlantic seaboard against marauding aircraft. On his first scramble in 1985, Hadfield and his mates turned away a Soviet Bear bomber that had entered Canadian airspace to probe the country's defences. Soon, Hadfield impressed his American colleagues in competitions between Canadian and American F-18 pilots.

Hadfield was accepted for the prestigious U.S. Air Force Test Pilot Training School at Edwards Air Force

Base in California, where many U.S. astronauts tested aircraft before joining NASA. Hadfield finished first in his class and set to work on his own test-flying program to learn more about a troubling problem with the F-18. His work on the control problem, which has been credited with saving lives, also earned Hadfield the title of U.S. Navy Test Pilot of the Year, the first time a foreign pilot had won the accolade.

Hadfield didn't consider himself ready when Canada sought its first astronaut group in 1983, but he was ready in 1992 when the next opportunity opened up. Hadfield, his wife Helene, their two boys and their daughter moved to Houston, where the Canadian became a popular member of NASA's astronaut group. Having grown up in a musical home, Hadfield loves to play guitar, and he was soon drafted to join the NASA astronaut band, Max Q, which plays at astronaut and charity events around Houston. When in Canada, he performs folk music with his brother Dave, who writes and records music when he isn't flying.

After a launch planned for the previous day was scrubbed because of bad weather, Major Hadfield and the other four astronauts of STS-74 blasted off through low clouds for Mir early on the morning of November 12, 1995. "It was just a tremendous feeling to be actually doing something that I've dreamed about doing for nearly 30 years," Hadfield said later.

The 5.1-metre-long docking module was stowed in *Atlantis'* payload bay. The day after launch, Hadfield was entrusted with moving the module into place to

join it with Mir. The orange-coloured module was originally been built in Russia for use by the Soviet space
shuttle Buran. But Buran flew only once, in 1989, and
then was abandoned because of its huge cost. When
Russia and the U.S. agreed on joint flights to Mir, the
module was flown to the U.S. to be refitted.

The shuttle's Orbiter Docking System also sat in
Atlantis' payload bay. On day three of the flight, Hadfield became the first Canadian astronaut to operate
the Canadarm in space when he used it to grab the
docking module and lift it out of its cradle in the payload bay. Hadfield turned the docking module around
and positioned it directly above the Orbiter Docking
System. While Hadfield was directing these precision
maneuvers, he got help from astronaut Bill MacArthur,
who was running an advanced version of the Space
Vision System. The new system incorporated improvements based on the results of Steve MacLean's work
with the SVS on his mission three years earlier.

Finally, the petals of the module and the shuttle's
docking systems were in perfect alignment. The module sat just 15 centimetres above the shuttle's docking
ring. The Canadarm didn't have enough strength to
join the module to the docking system, so Hadfield let
the arm go limp just as commander Ken Cameron
fired six shuttle thrusters, joining the module to the
Orbiter Docking System. The docking module
protruded from the shuttle payload bay, firmly joined
to *Atlantis*. "It was letter perfect," Hadfield said. A day
later, *Atlantis* reached Mir, and Cameron joined the

Canadian astronaut Chris Hadfield during his spacewalk on the STS-100 mission in 2001

other end of the docking module to Mir's Kristall module in a precision docking operation.

When the shuttle crew passed through the docking module to Mir, they brought fresh food, mail and gifts for Mir's three-member crew. Hadfield brought maple sugar candies, and more memorably, a lightweight guitar that was used by Mir crews for the rest of the station's life. For the next three days, Hadfield and his crewmates visited Mir, transferring precious water to the station along with vital equipment. New solar arrays were attached to the docking module. When Cameron pulled *Atlantis* away, he left the docking module attached to Mir. After eight days in orbit, the shuttle returned safely to the Kennedy Space Center, wrapping up the mission and opening the door to regular visits to Mir by the shuttle. A NASA history of the Shuttle–Mir program notes: "STS-74 was the first space shuttle mission to actually help build a space station."

Hadfield said his three days in Mir gave him "a succession of a million moments" to remember. While many of those moments concerned the joys and successes of the mission, Hadfield and his colleagues found a crowded station filled with aging and broken-down equipment. Starting a few months after STS-74, six American astronauts made long term stays of several months each on Mir. One of the Americans witnessed a fire when an oxygen generator blew out of control, and another American and his two Russian colleagues survived a cargo ship's collision with Mir that punched a hole in the station.

In spite of these problems, the Americans and Russians learned an important set of lessons that prepared the way for the International Space Station. The shuttle stopped visiting Mir in 1998, but the station continued operating until it was destroyed in a controlled re-entry over the Pacific Ocean in 2001, after 15 years in orbit. Chris Hadfield knew better than most the changes the end of the Cold War has caused, having defended his country against a Soviet bomber in 1985 and becoming an honoured guest on a Russian space station a decade later.

CHAPTER FIFTEEN

Gaining Experience in Space

WHILE SHUTTLES WERE VISITING MIR IN THE LATE 1990s, other shuttle missions were testing new equipment and carrying out experiments in anticipation of the International Space Station. Four of those flights carried Canadian astronauts, preparing them for the important work ahead on the station.

Marc Garneau got his second ride aboard the shuttle *Endeavour* on May 19, 1996, when he and five other astronauts flew on STS-77. In the 11 years since his first mission, Garneau had made important changes in his life. This time, Garneau was a full member of the crew as a mission specialist. When he began his mission specialist training in 1992, he married Pamela Soame of Ottawa. The couple and Garneau's two children from his previous marriage moved to Houston, where he completed his mission specialist training and served as a capsule communicator while he prepared for his second trip into space.

The astronauts released a *Spartan* satellite equipped with a 15-metre-wide inflatable antenna that was to test the utility of future inflatable structures in space. During the flight of the *Spartan*, the crew sent back spectacular television views of the Mylar antenna

Canadian astronaut Marc Garneau on board the shuttle in orbit

~⁂~

unfolding and inflating its 10-storey-high struts and
of the antenna's behaviour once it was fully deployed.
Previous inflatable structures have sometimes failed
to inflate properly, but the antenna worked nearly
flawlessly. Controllers were surprised by mild tumbling
and ripples moving across the silvery face of the structure. After the antenna separated from the *Spartan*,
Garneau retrieved the satellite with the Canadarm
and returned it to *Endeavour*'s payload bay.

Endeavour also carried a small Spacelab module in
the payload bay, which allowed the crew to carry out
a set of experiments. These included more materials
testing taking advantage of the weightlessness of space
and the Canadian-built Aquatic Research Facility to

examine the effects of weightlessness on life forms. Other experiments included the Fluids Generic Bio-processing Apparatus, better known to the public as an outer space dispenser for Coca-Cola. The device mixes carbon dioxide and the fluid portion of the carbonated beverage for taste tests by the astronauts.

During the flight, Garneau and his crewmates kept in touch with Earth through CAPCOM Chris Hadfield at Mission Control in Houston. When *Endeavour* returned to Cape Canaveral after 10 days in space, Garneau became the first Canadian to experience re-entry from the flight deck, with its many windows providing breathtaking views of the flames enveloping the shuttle as it slowed down in the atmosphere on its way to a landing.

Barely three weeks after the completion of Garneau's second flight, his backup from his first flight in 1984 finally got to go into space. Robert Thirsk was launched aboard *Columbia* as part of the Life and Microgravity Spacelab flight on June 20.

Thirsk was born in New Westminster, BC, in 1953 and grew up in various locations in BC, Alberta and Manitoba, all of which claimed him as a native son when he went into space. After earning his bachelor's degree from the University of Calgary, he earned a master's degree in engineering from the prestigious Massachusetts Institute of Technology and a medical degree from McGill University in Montréal. The following year, he joined the Canadian Astronaut Program. While he waited for his first flight, Thirsk kept busy

Canadian astronaut Robert Thirsk in space during the flight of STS-78

~∂Ҳ∂~

conducting research on the effects of weightlessness on the cardiovascular system. He commanded a seven-day simulated shuttle mission in 1994 in a hyperbaric chamber in Toronto with fellow Canadian astronauts Julie Payette, Dave Williams and Mike McKay. Thirsk also worked with educators to get Canadian students involved in space experiments.

Thirsk's flight involved seven astronauts who used a Spacelab module like the one used in Bondar's flight, although this module was outfitted with a new set of experiments. A number of experiments involved new investigations of the effects of weightlessness on astronauts and other organisms. One notable effort designed by Douglas Watt of McGill was the Torsion Rotation Experiment that tested how head movements

affect motion sickness. In the experiment, the astronauts did not move their heads without moving their
torsos, a strategy some astronauts have used to avoid
space sickness. Other medical tests included special
goggles that displayed wavy lines to track eye movements in weightlessness. Yet another experiment
probed the function of leg muscles in space.

The crew of STS-78 ran a battery of materials pro-
cessing experiments, examined fluid flow and grew
crystals in weightlessness. *Columbia* also carried the
Shuttle Amateur Radio Experiment (SAREX) that
allowed the astronauts to talk to ham radio enthusiasts on the ground.

Thirsk's flight was also notable for symbolic reasons.
The crew patch marking the mission depicted *Columbia*
in the style of West Coast Native art, as did Thirsk's
personal patch. The shuttle carried an Olympic torch
for the upcoming Atlanta games. Thirsk brought six
hockey pucks and the 1970 Stanley Cup ring belonging to his boyhood idol, Boston Bruins great Bobby
Orr. When *Columbia* returned to the Kennedy Space
Center on July 7, STS-78 went into the books as the
longest shuttle mission ever flown at nearly 17 days.
A few months later, another mission was flown.

Another Canadian astronaut with BC roots took to
the skies 13 months later aboard the shuttle *Discovery*
on mission STS-85. Bjarni Tryggvason, the last of the
original six astronauts to get a flight into space, ended
more than 13 years of waiting when *Discovery* lifted
off on August 7, 1997. "Although it's been a long

Canadian astronaut Bjarni Tryggvason on board *Discovery* during the
flight of STS-85

time, I never thought I wasn't going to get that
chance," he said shortly before his flight. "If I had
been seriously worried about that, I would have gone
back to doing university research." Tryggvason, 51 at
the time of his launch, was born in Reykjavik, Iceland,
but was raised mainly in Nova Scotia and Vancouver.
He got his engineering degree from the University of

British Columbia, followed by graduate engineering studies in applied mathematics and fluid dynamics at the University of Western Ontario. An accomplished pilot, Tryggvason was working as a researcher at NRC when he joined the astronaut program.

While he did a great deal of work developing the Space Vision System, Tryggvason is best known for helping develop the Microgravity Isolation Mount (MIM). This device is needed because the weightless (or microgravity) environment inside most spacecraft is not perfect, usually because of the movement of thruster rockets or machines and astronauts inside. The creation of crystals in space can be disrupted if there is any movement in the spacecraft. Experiments that require perfect microgravity conditions are placed in the MIM, where they are sheltered from movements that perturb the gravitational environment. Even before a MIM was flown on STS-85, the device was tested on NASA aircraft that provide brief periods of weightlessness, and more importantly, on board the Russian Mir space station. Findings from these flights are being used to develop similar equipment for the International Space Station.

During their 12 days in space, the six astronauts deployed and captured a satellite that made astronomical observations and used special instruments in the payload bay to make observations of the Sun. The crew also tested a small Japanese robot arm similar to one that will be used on the Japanese Experimental Module on the International Space Station.

A few months later, on April 17, 1998, Dave Williams flew into space as part of the crew of the final Spacelab mission. The flight, known as Neurolab or STS-90, had a crew of seven astronauts on *Columbia*. The crew carried out 26 experiments, including two from Canada, on how humans and other animals respond to weightlessness. This time the emphasis was on the neurological system. Many experiments, including the Canadian experiments, dealt with motion sickness. One examined when and how visual cues take over from other cues as astronauts adapt to the loss of gravity. Rats, mice, snails and fish were on board the Spacelab, requiring a veterinarian. Even cellular biology experiments flew on board the shuttle.

The astronauts also became guinea pigs when they rode a rotating chair to see how they responded to the sensory cues. They also wore equipment while they slept to record how conditions in space affected their sleep. The hope is that the experiment and the lightweight experimental apparatus will benefit insomniacs. The set of experiments on Neurolab was also aimed at helping people back on Earth with disorders such as Parkinson's disease.

Williams and his STS-90 crewmates spoke briefly during their flight to Dave Thomas, the final American astronaut to live on board Mir. Chris Hadfield, who worked as a CAPCOM during STS-90, found that the cluttered interior of the Neurolab he saw on television was similar to the mess he saw inside Mir. "This is not because it is inefficient; that's how life in zero-G will always be. It is not necessary to put things away, or on shelves, or in cupboards when there's no gravity and no company coming by. Never believe a pristine, open mock-up or simulation; weightless space stations will always look like a yard sale," Hadfield said.

Williams was the medical officer on the flight, which lasted just under 16 days. He was also the flight engineer during the ascent, thanks in part to his experience as a pilot. He learned to fly with lessons from his wife, Cathy Fraser, an Air Canada pilot who strongly urged Williams to apply to the astronaut program.

Williams was born in 1954 in Saskatoon and attended high school in Beaconsfield, Québec. Although Williams followed the space program in his youth, he thought he would find adventure as a scuba diver and a marine biologist. He obtained three degrees at McGill, including a bachelor and a master of science in biology. His research awakened an interest in human biology, so Williams earned a medical degree at McGill. He worked as an emergency room physician in Toronto and Waterloo and was acting director of emergency services at Sunnybrook Medical Centre in Toronto when he was selected to join the astronaut program in 1992. Three years later, he began training in Houston to become a mission specialist. When Williams returned from STS-90, he served for four years as head of the Space and Life Sciences Directorate at NASA's Johnson Space Center in Houston, making him the first non-American to hold such a senior position in NASA.

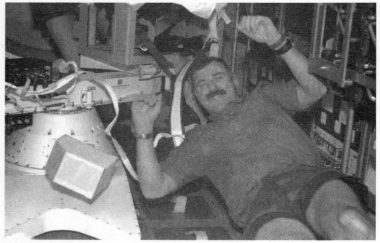

Canadian astronaut Dave Williams in space during the flight of STS-90

~∞∞~

In 2001, Williams' interest in underwater work became a reality when he took part in a seven-day mission in an undersea laboratory off the Florida Keys that tested remote surgical techniques for future space flights and underwater exploration. Williams thus became the first Canadian to live and work both in space and under the ocean. He returned to the undersea lab in 2006 to command an 18-day mission. The techniques tested included telerobotic surgery and remote consultations with expert physicians.

The end of the STS-90 mission meant that the shuttle's days as the primary experiment platform for space flight were coming to an end. The Shuttle–Mir program was also ending, and NASA was clearing the decks for the International Space Station. Soon, most shuttle missions would be dedicated to building the new station. Those flights could not go ahead without Russian components launched from Kazakhstan, and Canadian help would be needed from the start to put the new station together.

CHAPTER SIXTEEN

The International Space Station

THE SAGA OF THE INTERNATIONAL SPACE STATION BEGAN with a late morning launch on November 20, 1998, from the desolate and dusty Baikonur space centre in Kazakhstan. A Proton rocket carried the first station module into orbit. Inside a shroud atop the rocket was the Zarya control module, the first part of the station to actually enter space. Zarya, named for the Russian word for sunrise, entered orbit and opened its solar panels shortly. The event was witnessed by top space officials from all the nations taking part in the ISS. The Canadian delegation was headed by Canadian Space Agency President Mac Evans, who had worked hard through the previous 15 years to make the station a reality, and astronaut Julie Payette, who was already training to be one of the first visitors to Zarya. "It's like a birthday party," a jubilant Evans said at the celebration following the historic launch.

Two weeks later, *Endeavour* roared off its pad at Cape Canaveral carrying a crew that began the lengthy task of building the ISS. Six Americans and a Russian on the crew of STS-88 would need the Canadarm and the Space Vision System to do their job. In orbit, astronaut Nancy Currie lifted another module for the station—a connecting hub called Unity—out of the payload bay. In an operation similar to Chris Hadfield's work with Mir's docking module three years earlier, the Canadarm moved Unity precisely to a spot above the Orbiter Docking System. Shuttle commander Bob Cabana fired *Endeavour*'s thrusters to join the 12-tonne Unity to the docking system. A day later, the shuttle reached the Zarya module. Soon, Currie grappled Zarya with the Canadarm. To join Zarya to Unity, she would depend on television cameras and the Space Vision System because the large Unity module blocked her view of Zarya. For more than two hours, Currie worked to line up the docking systems of the two modules. Her work was slowed by a minor problem with the vision system, but finally, the two modules were aligned perfectly. Once the Canadarm's grip was loosened, Cabana drove the two modules together, creating a structure that extended seven stories out of *Endeavour*'s payload bay. Spacewalkers from *Endeavour* connected systems from the two modules, and the shuttle's astronauts became the first visitors to the ISS before they headed home.

The station was left to fly on its own for nearly six months until the launch of *Discovery* on STS-96. Early in the morning of May 27, 1999, Julie Payette, Russian cosmonaut Valery Tokarev and five Americans lifted off from Pad 39B at the Kennedy Space Center, bound for the ISS. *Discovery* spent nearly two days chasing the ISS. During that time Payette and fellow astronaut Ellen Ochoa tested the Canadarm in preparation for their work later in the mission. A day after the shuttle docked with the station, Tamara Jernigan and Daniel T. Barry conducted a spacewalk that lasted nearly eight hours. The pair moved tools and two cranes from the shuttle's payload bay to the external surface of the station. During the walk, Payette worked inside the shuttle as the spacewalk coordinator or "choreographer." This involved communicating with her two spacewalking crewmates while they were outside, keeping them on schedule and helping them into and out of their suits before and after the walk.

The next day, Payette and her colleagues opened the hatches between *Discovery* and the station. For three days, the astronauts moved three tonnes of equipment from the shuttle into the station's Unity and Zarya modules. Some equipment, such as mufflers to reduce the volume of Zarya's life support systems, required installation. Other things, including clothing, sleeping bags, water, computers and cameras, were stowed for later use. Some of Zarya's batteries had been operating erratically since launch. Payette and Tokarev had to remove 12 control modules for four of the six batteries and replace them with fresh controllers carried aboard the shuttle. The operation involved removing floor panels and changing out connectors, a complicated task that required Payette to obtain special training in Russia. Payette also operated the Canadarm to conduct an inspection of dots placed on the outside of both station modules for use by the Space Vision System.

Julie Payette was 35 when she became the first Canadian to visit the ISS and the eighth Canadian astronaut to fly in space. Before she joined the

Canadian astronaut Julie Payette

astronaut program in 1992, Payette worked at Bell Northern Research on a project to develop computers to use human language and voice recognition and was a visiting scientist at IBM's research laboratory in Zurich. After she attended school in Montréal and at the United World College of the Atlantic in Wales, Payette earned engineering degrees from McGill University and the University of Toronto. Payette is active in sports, including triathlon, skiing, racquet sports and hiking. She also plays the piano and has sung with Orpheus Singers, the Tafelmusik Baroque Orchestra Choir in Toronto and the Montréal Symphonic Orchestra Choir. After becoming an astronaut she earned her pilot's license, and she has since learned to fly military jet aircraft. She also learned deep sea diving and added Russian to the other languages she can speak—French, English, Italian, German and Spanish. In 1996, Payette moved to Houston to begin her mission specialist training in preparation for her flight assignment.

"I often say I wanted to become an astronaut like others wanted to become a ballerina or a fireman," she told the media when she was selected. She admitted to being "flabbergasted" at being chosen, because she was only 28 at the time. Payette, who is married and since her flight has had two children, carefully guards her private life. "This is one of the not-so-perfect aspects of the job," she said of the travel and the high visibility affecting her family life.

After five days joined to the station, *Discovery* undocked and conducted an inspection of the ISS's exterior. Later on, Payette deployed a spherical satellite covered with 900 small mirrors, known as *Starshine*. Students helped build the satellite by preparing the small mirrors, and during *Starshine*'s lifetime, more than 25,000 students from 18 countries tracked its flight. After nine days and 19 hours in space, *Discovery* glided to a night-time landing near its launch site at the Kennedy Space Center.

The ISS needed a great deal of work before astronauts could live on board. Much of that work waited while the Russians dealt with delays with the Zvezda service module. The Americans also postponed shuttle launches when the flight following Payette's STS-96 experienced problems during launch. Finally, in July 2000, Zvezda, which at one time was to be Russia's follow-up to Mir, was launched from Baikonur. The Russian space agency's need for financial help was underscored by the Pizza Hut logo painted on the side of the Proton launch rocket in exchange for a fee of $1 million USD. Zvezda, Russian for "star," contained living quarters for the ISS crew and important life support systems for the station. During the first few months of the year 2000, three shuttle flights visited the station, carrying equipment for installation inside the station and on its exterior.

Finally, the station's first resident crew, Expedition One, lifted off from Baikonur in a Soyuz ferry craft on October 31, 2000. Russian Cosmonauts Yuri Gidzenko and Sergei Krikalev and U.S. astronaut Bill Shepherd arrived at the ISS on November 2 and began the period of continuous occupation on ISS that continues to this day.

On the evening of November 30, Expedition One's first guests lifted off from Cape Canaveral. The crew of STS-97 on the shuttle *Endeavour* included four American astronauts and Canadian Marc Garneau, who was making his third flight into space. On the third day, *Endeavour* docked with the ISS. The veteran Garneau assisted the commander and pilot during the intricate operation. But the Expedition One crew had to wait to welcome its visitors because much of the work of STS-97 involved walking in space. Until the space-walks were complete, the atmosphere inside the shuttle had to be kept at a lower pressure than the station, so the hatches between the shuttle and the station remained sealed. Joseph Tanner and Carlos Noriega made three spacewalks. Garneau remained inside *Endeavour* and acted as spacewalk coordinator, but he was also trained to go out and walk in space in case one of the designated spacewalkers wasn't able to do the job. The main task of the mission was to install the first set of American solar arrays on the station. Although the two Russian modules had their own solar arrays to collect power, the station required the large U.S. arrays to operate properly. Several missions to install more solar arrays were scheduled to follow STS-97. Garneau used the Canadarm to move the solar panels and the truss to which they were attached out of the cargo bay and into a position near another truss on the station that had been delivered on an earlier flight. Like astronauts on other missions that were delivering equipment to the ISS during this time, Garneau used the Space Vision System for this operation.

The next day, Tanner and Noriega moved outside, and Garneau used the Canadarm to move the truss to its attach point on the station. The two spacewalking astronauts then bolted the truss to the station and attached power and data cables. Once the connections were made, the astronauts gave computer commands to the 34-metre-long solar panels to unfurl. Only one of the two panels deployed correctly, but the balky

Underwater training for spacewalks

panel deployed the next day when they tried again.
Two days after the first spacewalk, Tanner and Noriega
went outside again, connected more cables to the
solar panels and did some work on a docking port.

On the third spacewalk, two days later, the two astronauts completed their scheduled tasks and even did some work originally scheduled for a later mission.

When buildings are topped off, workers customarily place an evergreen tree on top to signify the accomplishment. For the space station, the spacewalkers attached an image of an evergreen tree to the top of the solar panels. During all three spacewalks, Garneau supervised their activities and kept an eye on the clock to make sure that they weren't overstressing their life support systems, which spacewalking astronauts carry on their backs. "You may have to re-jig the [spacewalk] and reassign priorities—always keeping in mind that the first priority is their safety," Garneau said. "That's where you earn your money."

Six days after the shuttle arrived at the ISS, the spacewalks were completed, the pressure differentials between the two spacecraft were reduced, and the five *Endeavour* astronauts visited the Expedition One crew in the ISS. During the 24 hours that the hatches were open, the crews exchanged equipment, held a press conference and had a brief visit.

Inside the station, astronauts move along modules that are placed in line with each other. At one end of the station is the Zvezda service module that holds the astronauts' living quarters. It includes two closet-like bedrooms where astronauts can rig their sleeping bags. These keep astronauts in place while they sleep in a weightless environment. Each bedroom is decorated with family photos and other keepsakes belonging to the astronauts, who live on the station for six-month stretches. When long-term crews have three members, the third member must find a sleeping spot elsewhere in the station, such as an airlock module. When astronauts visit from the shuttle, as Garneau and his colleagues did, they sleep on board the shuttle. During visits, astronauts from the station and the shuttle often gather for meals in Zvezda. The station also contains exercise facilities and a bathroom facility that uses suction to remove waste. It is important for astronauts to exercise because legs are not used much normally in weightlessness. One form of exercise is running on a treadmill where astronauts strap themselves down so they don't float off when they are running.

Next to Zvezda is the Zarya control module, which houses the machinery that helps keep the station going. Zarya opens into a large space called the Unity node, which since 2001 has been joined to the Destiny module and the Quest airlock. More modules will be joined to the Unity node in the future. Astronauts can also visit the Soyuz ferry spacecraft that is always docked to the station to allow astronauts to leave at any time should there be a problem.

When the Destiny laboratory module was added to the station in 2001, it became a popular gathering spot because it includes a large window that is ideal to enjoy views of the Earth passing by. Also in 2001, the station was enlarged with the addition of the Russian airlock module Pirs and the American airlock module Quest. Station crews can go out for spacewalks using either airlock module, wearing either Russian or American space suits. Outside the station, the modules are dwarfed by the large solar panels that collect power for the station, a set of radiators that eliminate heat built up by the machinery and trusses that are attached to the panels and the radiators. The station always impresses crews when they arrive or leave. Garneau's flight to the ISS was no exception.

Two days after leaving the station, nearly 11 days after launch, *Endeavour* landed safely on the shuttle landing strip at Cape Canaveral. Shortly after landing, Garneau announced that STS-97 was his last flight. The following February, he became executive vice-president of the CSA, and in November 2001, a year after hanging up his space suit, Garneau succeeded Mac Evans as president of the agency.

CHAPTER SEVENTEEN

The Mobile Servicing System

WHILE MARC GARNEAU WAS WRAPPING UP HIS THIRD AND LAST space flight, NASA and the CSA were fully engaged in final preparations to install Canada's contribution to the International Space Station (ISS). Work on that contribution, known as the Mobile Servicing System (MSS), had been underway since Canada joined the station program in 1986. During that time, the MSS had undergone many changes as the station evolved from the Space Station Freedom design of the 1980s to the International Space Station that began flying in 1998. The station's management and financial woes, plus the addition of the Russians onto the project, required major redesigns for the MSS. When the Canadian government nearly backed out of the ISS in 1994, it insisted on cutbacks to Canadian costs as the price of staying in the program. Even the prime contractor of the MSS faced big changes. In 1999, Spar Aerospace decided to get out of the space business and concentrate on aircraft maintenance. Spar sold its robotics division in Brampton, Ontario, to MacDonald Dettwiler and Associates (MDA) of Richmond, BC, a company that has a long history in the space business. The operation was rebranded as MD Robotics.

The $1.4 billion MSS is made up of three parts—
Canadarm2, the Mobile Base System and Dextre—each
of which posed major challenges to the Canadian
team that built them. The best-known part of the MSS
is Canadarm2, which at first glance is similar to the
shuttle Canadarm. In reality, it is far more advanced.
Officially known as the Space Station Remote Manipulator System (SSRMS), Canadarm2 is not fixed to
a single point like the shuttle arm. Because it is needed
at various points around the station, Canadarm2 can
move end over end like an inchworm or a slinky. To do
this, the arm needs far more than the ability to grab
things at both its ends. It needs to be able to maintain
its electrical and communications connections with the
station while it moves.

The shuttle Canadarm is required to operate for only
the length of each shuttle mission, usually two weeks
or less, before it returns to Earth for maintenance.
Once Canadarm2 is launched, it is in space for its entire
lifetime. Each part and system must be able to be
repaired or changed out in space. The shuttle arm has
six degrees of freedom: it has two joints in its shoulder, one joint in its elbow and three joints in its wrist.
Canadarm2 has an extra joint in its shoulder and can
move around much farther than the shuttle arm,
which has limited movement in its elbow joint. The
15-metre-long shuttle Canadarm weighs 410 kilograms; the 17-metre-long Canadarm2 weighs 1800
kilograms and can move much heavier loads than its
shuttle counterpart. Canadarm2 is also equipped with

four television cameras to allow astronauts and ground controllers to keep track of its work.

One of the major challenges in the Canadarm2 construction process was getting the arm from the ground to the ISS. The arm arrived at the Kennedy Space Center in 1999 to be prepared for launch. Canadarm2 was folded up and placed inside a pallet that would protect it from the buffeting of a shuttle launch. The installation of the arm was one of the most complicated tasks for the shuttle astronauts. That task fell to the crew of STS-100 and the shuttle *Endeavour*. The training process would be a long one.

A few weeks after he returned from his first space flight late in 1995, Chris Hadfield got a call asking if he would like to work on the flight that would install Canadarm2. Along with American Scott Parazynski, Hadfield would walk in space to install the arm on the station. With his eager agreement to join the mission, Hadfield began five years of hard training to prepare for STS-100. The training involved education on the systems inside the arm and simulations of work inside the shuttle, but above all, he underwent demanding physical training in space suits.

The challenges that outer space construction workers like Hadfield face are different from anything on Earth. They must take care to ensure that they are working from a firm base, because anyone even turning a wrench while floating in space will be flung around unless he or she is fastened down. Early astronauts found themselves exhausted by even the simplest tasks outside their spacecraft in orbit, and so they learned two things: they needed to be restrained when they worked, and they needed to practice hard before they went into space. Although some practice work is done on board the aircraft that provide brief 30-second periods of weightlessness and more recently using a virtual reality device, the best place to prepare for walking in space is in the water.

In preparation for the station program, NASA built a gigantic swimming pool at the Sonny Carter Training Facility near the Houston Space Center. The pool is 31 metres wide, 61 metres long, 12 metres deep, and contains more than 23 million litres of water. Astronauts wear their full space suits along with weights to keep them underwater. Specially trained personnel in scuba gear are always nearby in case something goes wrong. Inside the pool are full-scale replicas of the shuttle payload bay and space station modules.

While practicing the pool gives astronauts a good idea of what they will face in space, it is still hard work. Astronauts who don't trim their nails find that their fingers are bruised after a day in the pool. And when they hang upside down, gravity pushes their bodies down onto the shoulder joints of their space suits. Each simulation typically goes on for several hours.

While Hadfield and Parazynski prepared for their construction work, the rest of the crew of STS-100 was selected and trained. The seven-member crew included a Russian cosmonaut and a European astronaut from

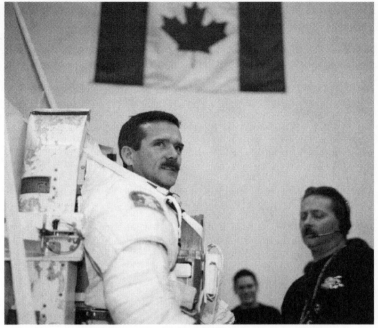

Canadian astronaut Chris Hadfield prepares for spacewalk training in Houston, Texas.

~oXo~

Italy. Canadarm2 and the rest of the shuttle's cargo were prepared at the Cape. STS-100 was slated for launch in April 2001, five months after Garneau's flight to the ISS. Between the two flights, a shuttle mission delivered the American-build Destiny laboratory module to the ISS in February, and in March, another shuttle flight brought up the Expedition Two crew and carried home the Expedition One crew from the ISS. While docked to the station, the shuttle crew brought equipment to the ISS, including a platform to mount Canadarm2 on the station.

Finally, the lengthy training schedule for STS-100 came to an end. *Endeavour* stood on the pad, ready to receive its crew. Right on schedule, in the early afternoon of April 19, 2001, the shuttle lifted off on its lengthy chase to catch up with the station. Near the end of the second day, *Endeavour* docked with the station, but because of the pressure differences between the two spacecraft, the crews would not meet until later. The following morning, on April 22, Hadfield and Parazynski donned their space suits and stepped out into the vacuum of space at 6:45 AM eastern time. For the first time, an astronaut wearing the Maple Leaf on his suit emerged into space. "Oh man, what a view!" Hadfield, now a colonel, exclaimed as he made history.

After the two spacewalkers installed a new antenna on the station, they turned to their main task, installing Canadarm2. Astronaut Jeff Ashby had already used the shuttle Canadarm to lift the pallet containing Canadarm2 out of the shuttle payload bay and attached it to the Destiny module on the station. For much of the walk, Hadfield worked with his feet attached to a platform at the end of the Canadarm. Hadfield and Parazynski connected cables between the station and Canadarm2 and removed insulating blankets from the arm along with the 32 bolts and eight "superbolts" that secured it in place during launch. Then Ashby maneuvered the shuttle arm with Hadfield's help to a position where they could unfold Canadarm2. The two spacewalkers installed and tightened eight fasteners that held the main booms of

Canadarm2 together. This operation ran into trouble when the fasteners refused to tighten properly. Mission control told the spacewalkers to turn them by hand instead of with a power wrench, and the fasteners finally tightened.

Aside from the balky fasteners, the two astronauts remarked that the spacewalk was going more smoothly than most simulations on the ground. But midway through the spacewalk, Hadfield noticed that something in his helmet, probably some defogging solution from his visor, was irritating his eyes. Closing his eyes and using a device that blows air into the helmet, he eventually cleared up the problem that could have shortened the spacewalk.

During the 7 hours and 10 minutes the two spent outside, they were able to pause and take in the view. During each 90-minute orbit, the spacecraft and astronauts spent half their time in the shadow of night and the remaining 45 minutes in daylight. On a night-time pass, Hadfield became quiet. "Oh wow. Northern lights," he finally said. His crewmates told him they were over Australia. "Southern lights, then. Wow, is that beautiful! With lightning down below me and the sky just lighting up." Then Hadfield shut off the lights on his helmet so he could enjoy the stars. "Oh wow." After more wows, Hadfield went back to work. At another point, Hadfield said to Parazynski: "Scott, when I was a little kid wanting to grow up to be an astronaut, this is what I wanted to do."

Back on the ground in mission control, flight controllers decorated their consoles with Canadian flags and wore red and white. Canadian astronaut Steve MacLean sat at the CAPCOM console wearing a Team Canada hockey jersey. "Chris, I think you'll like this," MacLean said, and then piped up to the shuttle a version of "O Canada" as sung by Roger Doucet, who was well known for singing the national anthem at the Montréal Forum before Canadiens games. MacLean pointed out that the space station was passing over Newfoundland, and Hadfield exclaimed that he could see the Avalon Penninsula from his cosmic perch. "We're really proud of your work getting Canadarm2 operational," MacLean added. "Scott and I were just the deliverymen, and it opens the door to everything we can do internationally," Hadfield replied.

Just as the spacewalk ended, the station crew commanded Canadarm2's first movement, and the next day, they "walked" Canadarm2 off the pallet to an attach point on the station. That day, the two crews met briefly, and the shuttle arm transferred an Italian module, Raffaello, a space-age moving van, to the station to transfer equipment to the shuttle.

The next day, two days after their first spacewalk, Hadfield and Parazynski floated out from the shuttle airlock again. During their 7 hours and 40 minutes outside, the two astronauts worked as electricians, making more connections between Canadarm2 and its attach points on the station. When one circuit failed, the astronauts succeeded in rewiring another. The spacewalkers also rewired an antenna and transferred a switching unit from the shuttle to the station. They were so successful that a proposed third spacewalk was cancelled because it was not needed.

The STS-100 crew resumed transferring equipment between *Endeavour* and the station the next day, but computer trouble on the station held up more activities with Canadarm2. Because of the problems, mission control granted the two crews two extra days together. Finally, three days late, on April 28, Canadarm2 handed over the pallet that contained it during launch to the shuttle's Canadarm. Once the historic "handshake in space" between the two Canadarms was completed, the pallet was stowed back in the shuttle payload bay, along with the Raffaello module that was filled with equipment to be returned to Earth. A day later, *Endeavour* and its crew separated from the station and its three-member crew. Two days after that, on May 1, *Endeavour* landed at Edwards Air Force Base in California—rain at Cape Canaveral prevented a landing there. The 11-day mission had been a complete success.

When *Atlantis*, the next shuttle to visit the station, arrived in July, Canadarm2 was put to work moving the station's new airlock, Quest, from the shuttle's payload bay to its location on board the station. Construction of the station continued into 2002. In June, *Endeavour* returned to the ISS on the STS-111 mission carrying the Mobile Base System (MBS), the second part of Canada's contribution to the station. The MBS is the work platform for both Canadarm2 and Dextre, and it serves as a storage facility for spacewalking astronauts. It contains four power sockets for the arm and rides on a small railcar that runs along a 109-metre track on the station's main truss. The truss was still under construction when the MBS was installed, and it required both the MBS and Canadarm2 to be completed.

When *Atlantis* delivered the MBS, two astronauts used Canadarm2 to attach the system to a small railcar that runs along a truss that was delivered to the station earlier in the year. Spacewalking astronauts made the connections between the MBS and the railcar and made repairs to a balky wrist joint on Canadarm2. After *Endeavour* departed, the station crew "walked" Canadarm2 off the Destiny module to its new main connect point on the MBS, and the arm continued to fulfil its vital role in station construction as shuttles delivered new parts for the station. Back on Earth, Canadian astronauts Steve MacLean and Dave Williams were named to missions scheduled to fly to the station in 2003. MacLean was training for spacewalks to help install trusses and solar panels on the ISS. And in 2004, the third part of the Canada's contribution to the station, Dextre or the "Canada Hand," was expected to be installed.

On January 16, *Columbia* lifted off on the first shuttle flight of 2003, carrying a crew of seven astronauts and a Spacelab module with experiments for a 16-day flight. STS-107 was the final shuttle flight of its kind. Scientific experiments were being shifted from dedicated shuttle flights to the space station.

When what had seemed to be a flawless flight returned to Earth on February 1, *Columbia* broke up over Texas. All seven astronauts perished. Investigations later showed that a piece of foam had fallen loose from the shuttle's external tank during launch and knocked a hole in *Columbia*'s wing. During the heat of re-entry, the wing disintegrated. The shuttle broke up under the stress.

While the investigation into the loss of *Columbia* went on, followed by changes to the shuttle system, NASA suspended shuttle flights. Crews continued to occupy the International Space Station, flying to and from the station aboard Russian Soyuz ferry vehicles.

Finally, in July 2005, a seven-member crew was ready to resume flying the shuttle. *Discovery* did not have any Canadians on board, but Canada assumed a prominent role in the flight. NASA placed restrictions on the shuttle and decided that on every flight, the bottom side of the vehicle would need to be inspected. NASA called on MD Robotics to build a boom to extend the reach of the shuttle Canadarm to make these inspections possible. The agency decided that the Orbiter Boom Extension System would fly on every shuttle mission to inspect the tiles and wing leading edges to ensure that there were no breakages like the one that doomed *Columbia*.

Discovery lifted off on July 26. The next day, the boom was deployed for the first time to inspect the shuttle's tiles. The boom carried a laser camera system developed by Neptec, an Ottawa company that also commercialized the technology for the Space Vision System. After the shuttle arrived at the ISS, *Discovery*'s mission was capped with a dramatic spacewalk when the station's Canadarm2 moved astronaut Steve Robinson under the shuttle so he could remove two protruding gap fillers that some feared could affect the tiles during *Discovery*'s re-entry. Although the mission ended safely, cameras caught a piece of foam falling off the external tank during the launch. Unlike the fatal *Columbia* mission, the foam did not strike the shuttle, but managers grounded the shuttle until the cause of the falling foam could be dealt with.

The next shuttle mission was postponed for nearly a year, but when it did fly in July 2006, it did not have any foam problems, opening the door for the long-awaited resumption of space station construction activities with Canadian Steve MacLean's flight aboard *Atlantis*.

After waiting to fly for more than three years due to the *Columbia* disaster, MacLean and his five fellow STS-115 astronauts also had to endure a further two weeks of launch postponements because of a lightning strike at the shuttle pad, a hurricane and technical glitches. *Atlantis* finally left the pad on September 9, 2006, and arrived at the ISS two days later. As soon as *Atlantis* arrived at the station, astronaut Dan Burbank picked up the 17-tonne truss from *Atlantis'* payload bay with the shuttle Canadarm. MacLean floated into the ISS and used the station's Canadarm2 to grab the truss from the shuttle arm and move it into place on the exterior of the station. With the move, which he

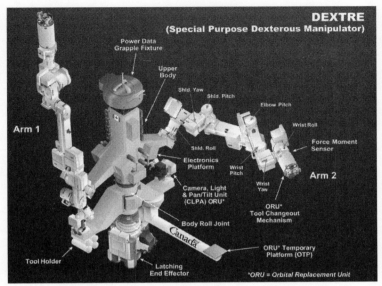

Schematic of Dextre, the Canadian-built Special Purpose Dexterous Manipulator

called "the great Canadian handshake," MacLean became the first Canadian astronaut to operate Canadarm2.

Four of the STS-115 astronauts carried out a series of three spacewalks to connect the truss to the station and remove bolts designed to protect the truss and its equipment during the launch. MacLean and space-walking partner Dan Burbank were responsible for the second spacewalk, where they completed the removal of bolts needed to free a new set of solar panels attached to the truss. During their seven-hour spacewalk, MacLean and Burbank dealt with a broken tool, a stubborn bolt that refused to budge until both of them worked together to free it and another bolt that floated off without being caught. "I hope they don't take it out of my wages," MacLean quipped. But the spacewalk was a success, and the two astronauts had time to do some work scheduled for the next spacewalk. The next day, the new solar panels, which will provide a quarter of the station's power, were unfurled. "It was an experience I wish everyone could share," MacLean said of his "quite amazing" experience as Canada's second spacewalker.

The flight of STS-115 marked the successful resumption of the complex work of space station construction and the return of Canadian astronauts to space after a five-year absence. Dave Williams is expected to be part of shuttle mission STS-118 in 2007 that will install another new truss on the station. The installation of new station modules and components, including Dextre, are scheduled for later missions.

Officially known as the Special Purpose Dextrous Manipulator, Dextre marks a major advance in space robotics. Instead of catching and moving large payloads like the two Canadarms, Dextre is designed to do fine work, such as changing out batteries, that at present only astronauts can perform. Although it is nicknamed the Canada Hand, Dextre resembles a torso with two arms attached. Each of the two arms has seven joints, and at the end of each arm is a mechanical hand that includes a set of retractable jaws able to grip objects and tools. Each hand is equipped with a retractable socket wrench, lights and a television camera. Dextre also has a sense of touch in that it can sense forces and torques in the payloads it is moving and compensate for changes in those forces. The robot is equipped with extra tools, lights and television cameras to help astronauts control the hand. This complex piece of technology can work on its own, but Dextre will usually be attached to Canadarm2.

The Mobile Servicing System also requires control and training facilities. While astronauts on board the station have actual control of the system, they come to the CSA headquarters on the ground in St. Hubert, Québec, to learn how to use the MSS. The training facility includes computer-based simulators and a virtual reality trainer that helps astronauts learn the three-dimensional environment outside the station that the MSS operates in. This training is especially critical because the astronauts' views of the MSS operations come through television cameras rather than with direct vision, as is usually the case with the shuttle Canadarm. CSA also has a Space Operations Control Centre to assist the NASA and Russian mission control rooms with the operation of the MSS.

In addition to building the station, Canada's scientists and astronauts have their share of the research being done on the station. Canadian astronauts are slated to spend months on board the ISS. Some of the Canadian experiments on the station include medical experiments and tests of equipment that exploit weightlessness to make new materials that can't be manufactured on Earth.

As construction goes on with the ISS, it will become a brighter star in the sky that can be seen when it passes over areas just after sunset or just before sunrise. The ISS can be seen as it passes over Canada as far north as Edmonton, and viewing times can be obtained from NASA's human spaceflight website or other websites. But even if the ISS is the most visible part of Canada's space activities, Canada's astronauts and space scientists are also looking in other directions for our future in space.

CHAPTER EIGHTEEN

On to Mars and Beyond

IN HIS 18 YEARS AS AN ASTRONAUT AND 4 YEARS AS PRESIDENT of the Canadian Space Agency, Marc Garneau was serious and determined about his work and the agency's mission of exploring the heavens. For example, when Canadian astronauts were grounded for six years following the 1986 *Challenger* disaster, Garneau was aware that the troubled space shuttle was not the only way to go into space.

Garneau made contact with representatives of the Soviet space program in 1987, when he attended a conference in Moscow and unexpectedly found himself invited to tour the cosmonaut training facilities at Star City, Russia, which was then still generally closed to Westerners. He kept in touch with his contacts in the Soviet Union, and by 1990, Garneau was hopeful that Canadians could fly on a Soviet spacecraft. "The Soviets have provided enough encouraging signs that we believe that if we were to go into negotiations with them, there would be very little standing in the way of organizing a mission," he said. But that effort ended with the fall of the Soviet Union in 1991.

Soon, the Russians became a partner in the International Space Station, but the CSA has continued to

quietly leave open the option of sending up Canadians on Russia's Soyuz spacecraft. After Chris Hadfield
returned from his triumphant visit to the ISS in 2001,
he was assigned to be NASA's chief of operations at
the Yuri Gagarin Cosmonaut Training Centre at Star
City. During his two years at Star City, Hadfield was in
charge of astronaut training for the ISS. Much of his
work involved negotiating with the Russians on ISS
crew activities. During his time in Russia, Hadfield
also learned to fly the Soyuz TMA ferry vehicle and to
walk in space using Russian space suits. By the time
Hadfield completed his two years, the shuttle was
grounded in the wake of the *Columbia* disaster, and
the only way to get to the station was by the Soyuz
TMA ferry—the latest generation of a spacecraft that
first flew in 1967.

In 2004, Robert Thirsk moved to Star City and
obtained his flight engineer certification for the Soyuz
TMA. He added to his training by serving as a backup
crewmember for European Space Agency astronaut
Roberto Vittori, who went to the ISS aboard *Soyuz
TMA-6* in April 2005. "Of course, after having completed this intensive training program, my desire to
someday work aboard the International Space Station
is greater than ever before," Thirsk said. "And it would
be terrific to fly up and down on the Soyuz spacecraft.
It is an amazing piece of technology."

Thirsk added to his experience by serving as a Crew
Interface Coordinator (the ESA's title for a capsule
communicator) in the ESA's new Columbus Control

Centre in Germany while his friend Vittori flew to the ISS and carried out experiments on board.

Today both Hadfield and Thirsk are preparing to spend several months on board the ISS as part of an expedition, not just a few days as a member of a visiting crew. It's possible that they may take a Soyuz to the station rather than the shuttle. And they or future Canadian astronauts may regularly use Russian or even future European spacecraft as their way into space.

The loss of *Columbia* and its seven astronauts in 2003 forced NASA and the U.S. government to recognize that the shuttle was nearing the end of its useful life. While the shuttle remains a technological marvel even 25 years after it first flew, it failed to significantly reduce the cost of sending payloads into space, one of the reasons the shuttles were built. Design compromises calculated to save money in the 1970s have come back to haunt the shuttle, along with features planned for military missions that never flew because of decisions made following the *Challenger* disaster.

Nearly a year after the loss of *Columbia*, President George W. Bush outlined a new vision of space exploration for NASA. That vision includes winding down the shuttle by 2010 and building a completely new spacecraft, the Crew Exploration Vehicle, to get American astronauts into space. The new vehicle returns to some of the simplicity and safety found in Apollo spacecraft of the 1960s and the Soyuz spacecraft of today. President Bush's vision also calls for a return to the Moon later in the second decade of the 21st century and then an expedition to Mars, although these plans can't really proceed until the Crew Exploration Vehicle is flying after 2010. NASA is designing the new vehicles, and although Bush has called for continued international cooperation in space, the place of Canada, Russia, Japan and the European Space Agency in the post-shuttle U.S. space program remains a big question mark.

China served notice of its bid to become a space power when it launched its first Taikonaut into orbit in 2003. While there is a good chance that Canada and the other ISS partners will continue to work closely with the Americans and get involved with their new spacecraft, they may choose—or be forced—to look to each other for continued access to space. That is why Canada's links with Russia and association with Europe, which goes back to the *Hermes* communications satellite in the 1970s, will become more important to Canada's space efforts as time goes on.

In 2002, Marc Garneau, then the president of the Canadian Space Agency, began speaking about the idea of a Canadian-built spacecraft exploring Mars. "This proposed mission would not be exclusively Canadian, for we recognize the value and the necessity, of collaborating with our international partners," he said in a speech about the idea. "However, it would be distinctly Canadian and would feature Canadian ideas, technologies and expertise." Garneau's idea was a bold departure for a space agency whose focus has been on earthly benefits. But the idea of Canadian technology going to Mars was nothing new. In a way, it was already old hat.

That's because Canadian equipment has already been to Mars, and work was already going on to enlarge that presence. When NASA's *Mars Pathfinder* landed on the Red Planet in July 1997, the *Sojourner* rover rolled off *Pathfinder* on ramps built by Astro Aerospace, then a subsidiary of Spar Aerospace, and used modems built by Dataradio of Montréal. *Sojourner* was the first roving vehicle to operate on Mars, and *Pathfinder* was the first lander on Mars since NASA's two Viking landers in 1976.

For many years, science fiction writers, astronomers and biologists have wondered whether the Red Planet harbours, or once harboured, any life. In the 19th century, some astronomers thought they saw canals, possibly created by intelligent beings. But the first space probes to pass by Mars in 1965 and 1969 showed a cold world without canals and with an extremely thin atmosphere that could not support life. The Vikings landed with equipment to search for evidence of life, but their findings seemed to point to a lifeless body. Interest in Mars returned in 1996 when scientists examining a meteorite from Mars found indications of life in Mars' past. *Mars Pathfinder* was the first of a number of spacecraft sent to try new ways of searching for life on Mars.

In 1998, the Japanese space agency launched a space probe, *Nozomi*, bound for orbit around Mars. The

The Japanese Mars probe *Nozomi*

~❦~

spacecraft carried the CSA's first venture in planetary science, an experiment from University of Calgary researchers called the Thermal Plasma Analyzer, which was designed to study the Martian atmosphere and its interaction with solar wind. *Nozomi* encountered a number of problems en route, including a partial failure of its burn to put it on a path for Mars and a later encounter with a solar flare. Finally, on December 10, 2003, Japanese space officials announced that *Nozomi* would not be able to enter Mars' orbit. The U.S. also lost two Mars spacecraft in 1999, but it succeeded in placing the *Mars Global Surveyor* into orbit in 1997, *Mars Odyssey* in 2001 and the *Mars Reconnaissance Orbiter* in 2006. Europe also put *Mars Express* into orbit around the red planet in 2003.

In January 2004, two American roving vehicles, *Spirit* and *Opportunity*, landed on opposite sides of Mars. Larger and more robust than the *Sojourner* rover, the vehicles continued exploring Mars more than two years after they landed. Both discovered evidence that water once flowed on Mars, and the presence of water indicates that living organisms could have once existed there. Both rovers carried image sensor chips developed by the Bromont, Québec division of the DALSA Corporation of Waterloo, Ontario.

NASA is preparing for the 2007 launch of the *Mars Phoenix* to a landing site near the planet's northern polar cap. *Phoenix* will use a robotic arm to dig a trench and retrieve samples for geological and chemical analysis. Instruments sensitive to minute quantities of organic molecules will help scientists assess the habitability of the icy layer for microbial life.

The CSA is making a major contribution to *Mars Phoenix* in the form of a weather station built by MD Robotics with Toronto's Optech Incorporated as a subcontractor. The Canadian weather sensing system will use a temperature instrument, an atmospheric pressure transducer and a lidar (laser radar) instrument to look at the exchange of water vapour between Mars' atmosphere and its polar caps, which contain both water ice and frozen carbon dioxide (dry ice). Researchers from York University and other Canadian universities are involved in developing and running the weather station on Mars.

Canada's space agency is looking to take part in future robotic Mars missions to be launched by NASA and by the European Space Agency. NASA is drawing up plans to send another rover to Mars, while the ESA is planning an ambitious orbiter and rover mission, known as Exo-Mars, and then a sample return mission as part of its Aurora program of Mars exploration missions.

In 2003, the federal cabinet turned down Garneau's ambitious proposal to take part in the 2009 NASA Mars Science Laboratory by contributing a drill and a radar device, but the next year he proposed that Canada build its own Mars robot spacecraft. "We do not want to be mere spectators but rather participants in the exploration of Mars," Garneau said.

His proposal has so far not got a response from the federal government, and late in 2005, Garneau left the CSA to pursue a career in politics. But the CSA is still pursuing the idea of sending Canadian robot spacecraft to Mars. In 2006, the agency announced that it is asking Canadian companies and researchers to study possible Canadian vehicles for experiments on Mars.

CHAPTER NINETEEN

Still Looking Up

WHILE CANADA'S SPACE PROGRAM HAS BEEN LOOKING FORWARD in the new millennium, it is also looking back to its roots: scientific satellites. In the summer of 2003, Canada launched two scientific satellites, its first since *ISIS 2* more than 30 years earlier.

SCISAT was launched on August 12, 2006 by a new type of booster rocket, a Pegasus, from an aircraft off the California coast. *SCISAT*, which was built by Bristol Aerospace of Winnipeg, carries instruments designed to learn more about the distribution of ozone in the Earth's atmosphere, especially in extreme latitudes. Chemicals added to Earth's atmosphere by humans have depleted ozone in the atmosphere, which reduces our protection from ultraviolet solar rays. The 150-kilogram *SCISAT* will help scientists assess measures being taken to protect the ozone layer and help countries decide on future environmental policy. The satellite is run by a team of scientists from Canadian universities along with other researchers from the U.S., Belgium, Japan, France and Sweden.

The experiments on *SCISAT* build on experience gained over the three decades since ISIS by Canadian scientists who built and launched experiments on satellites launched by other countries. For example, another Canadian ozone experiment flew on the Swedish *Odin* satellite launched in 2001. Sweden's *Viking* satellite carried Canadian experiments probing the auroras in the southern and northern polar regions of the upper atmosphere. Canadian experiments have also flown on board Europe's *Envisat* satellite and American environmental monitoring satellites such as *UARS*, *Terra* and *Cloudsat*.

The CSA is preparing a new and innovative satellite, *Cassiope*, for launch in 2007. *Cassiope* will combine a suite of scientific instruments studying the Earth's ionosphere with a telecommunications instrument designed to provide digital broadband courier service for commercial use by moving large digital files to almost anywhere in the world. This dual-use satellite will also be the product of collaboration between the government and private sector operators.

Canada's other scientific satellite launched in 2003 marked a deceptively modest step forward for Canada's original space explorers—its astronomers. On June 30, a Russian Rockot launch vehicle carried Canada's first space telescope into orbit from Russia's northern launch site at Plesetsk. While many people know about the *Hubble Space Telescope*, an 11-tonne satellite carrying a 2.4-metre telescope that has cost NASA billions of dollars, this satellite is almost the antithesis of *Hubble*.

Artist's conception of Canada's *MOST* (Microvariations and Oscillations of STars) satellite

❧

Known as *MOST* (Microvariability and Oscillations of STars) this satellite weighs only 60 kilograms, is the size and shape of a suitcase and carries a tiny 15-centimetre telescope. *MOST* cost only $10 million, a bargain for any sort of satellite. Many people call it the "Humble Space Telescope."

MOST does not send images back to Earth as *Hubble* does. Instead, it focuses on a given star, sometimes for weeks at a time, to allow its detectors to record even the tiniest changes in the amount of light emitted by the star, and radios the data about those light levels back to Earth. This is an effective way to look for changes in the star's surface that are caused by stellar versions of earthquakes. Seismologists use earthquakes to tell us about the nature of Earth's interior, and *MOST* aims to use similar methods to learn about the insides of distant stars. Stellar seismology is difficult to do from Earth because of the planet's day–night cycle, weather problems and other limitations on observing, but from a satellite like *MOST*, it's possible.

The idea for *MOST* originated with Toronto astronomer Slavek Rucinski, who wanted to investigate variable stars (stars which noticeably change brightness). Rucinski got in touch with CSA, which in the late 1990s had begun looking for proposals for small and inexpensive satellites, and a Toronto firm named Dynacon, where he got in touch with Kieran Carroll, a talented engineer and space enthusiast who put together a team to build a microsatellite that could meet *MOST*'s stringent pointing requirements. After the CSA approved the project, Rucinski had to leave because of other commitments. Jaymie Matthews, a colourful astronomer from the University of British Columbia, took over as chief scientist. Mathews is known for wearing a variety of t-shirts, including one showing cartoon character SpongeBob SquarePants, who resembles *MOST* with its square shape and two magnetometers. The *MOST* team also included the University of Toronto Institute of Aerospace Sciences (UTIAS) that worked with Dynacon to build the satellite, and the Centre for Research in Earth and Space Technology (CRESTech) in Ontario that worked on *MOST*'s telescope.

Once *MOST* was launched and tested, it was pointed at the star Procyon, thought to be similar to our own Sun with a similarly active seismology. The *MOST* scientists were stunned to get results that showed that Procyon was a stellar flatliner. Before releasing their results, they double-checked to make sure *MOST* was functioning properly. "Imagine if I were a doctor and the first patient I saw was normal in every way, except he had no heartbeat," Matthews said. The results have astronomers scrambling to revise their theories on how stars work.

Since then, *MOST* has been put to work looking for planets orbiting other stars, since planets exert small gravitational tugs on the stars they orbit, and these tugs can be picked up by the satellite. *MOST*'s detectors are so sensitive that they can even pick up light directly from some of these exoplanets. The *Spitzer* space telescope, a sister satellite to Hubble that observes in infrared light, narrowly beat out *MOST* for the title of first telescope to directly view an exoplanet. *MOST* has since come up with other unusual findings, including an example of a planetary tail wagging a stellar dog, where a planet is forcing its parent star to rotate in step with its orbit. It is also looking for Earth-sized planets orbiting other stars.

MOST is only the first Canadian astronomy microsatellite. Another satellite in this class, *NEOSSat* (Near Earth Object Surveillance Satellite), is being built to look for asteroids that could endanger Earth.

Canadian astronomers are also involved in international satellite efforts. While Canada is not a formal participant in the *Hubble Space Telescope*, Canadian astronomers do use the telescope and its data as part of their work. And when NASA considered for a time sending a robot to refurbish *Hubble* in place of the space shuttle after the loss of *Columbia*, the prime candidate was a variant of Canada's dexterous manipulator for the space station.

The CSA is also taking part in the *James Webb Space Telescope*, NASA's designated successor to *Hubble*, which is under development for a launch in the next decade. Canada also supplied sensors and other equipment to the *FUSE* (*Far Ultraviolet Spectroscopic Explorer*) satellite, which NASA launched in 1999 to look for the secrets of the Big Bang that scientists believe gave birth to the universe. Canadian astronomers are actively involved in *FUSE*'s research team.

To add to all that, Canada is involved in two upcoming astronomy satellites being readied for launch by the European Space Agency, including the *Herschel Space Observatory*, a telescope that will explore space in the infrared portion of the light spectrum, and *Planck Surveyor*, another ESA spacecraft that will look for information about cosmic background radiation and will help tell us about how our universe, our galaxy and the stars in it were formed. Canadian astronomers are also involved in an international effort that ties together radio telescopes on Earth with a Japanese satellite carrying a similar instrument to provide a clearer view of distant objects in the universe by using the data from the two with computers to create a virtual radio telescope effectively twice the size of the Earth.

These space projects are just a small part of the work being done by Canadian astronomers. Astronomy began growing in Canada after World War II, and this growth accelerated after *Sputnik* launched the space race in 1957. Many universities added astronomy departments during that time, and today many Canadian universities, from Memorial University in Newfoundland to the University of Victoria in BC, can boast world-class astronomy departments.

In the post-war years, the Dominion Observatory began meteor observing programs and also mapped geological features in Canada that were created by meteorite impacts, such as the New Québec Crater and Lake Manicouagan in Québec. Observing continued at the Dominion Astrophysical Observatory in Victoria and the David Dunlap Observatory north of Toronto. In the 1960s, many Canadian astronomers began to lobby for larger instruments. In the place of a planned telescope in southern BC, the country joined in the Canada-France-Hawaii Telescope, located at one of the world's best observing sites atop Mauna Kea, an extinct volcano in the island of Hawaii. In the 1960s, Canada entered the field of radio astronomy when observatories with large radio telescopes were built near Penticton, BC and in Algonquin Park in Ontario.

In 1970, government astronomy in Canada was moved under the wing of the National Research

Council of Canada. Five years later, the NRC consolidated its astronomy effort in the Herzberg Institute of Astrophysics, named after the Nobel Prize-winning Canadian physicist, chemist and astronomer Gerhard Herzberg. The institute operates the optical telescope in Victoria, the radio telescope near Penticton and Canada's share of telescopes outside the country, such as the twin 8-metre Gemini telescopes in Hawaii and the mountains of Chile, which are acknowledged as the Earth's best sites for optical telescopes.

Canadian astronomers are involved in many of today's important discoveries in astronomy, including those dealing with the origin of the universe, new moons around the outer planets of our solar system and asteroids that could pose a danger to Earth in the future, to name a few. In 1987, University of Toronto astronomer Ian Shelton discovered the brightest nova in 400 years. University of BC astronomer Gordon Walker and Bruce Campbell of the Herzberg Institute helped make the technical breakthroughs that allowed astronomers to begin the search for planets orbiting other stars. In 2005, Canadian astronomers could boast that their papers had the highest rate of citation by other astronomers in their scientific publications.

Nearly 2000 metres underground, in an abandoned mine near Sudbury, Ontario, is a giant tank of heavy water that can detect neutrinos (unusual particles emitted by the Sun and other stars). The Sudbury Neutrino Observatory is part of a worldwide search for information on these unusual particles. Canadians are also pioneering the construction of telescopes that use a giant rotating mirror of liquid mercury to replicate the strength of large telescopes at a fraction of their cost.

Like the Canadian space program, Canadian astronomers have helped lead the world despite receiving government support that is a fraction of spending in other countries, even when adjusted for Canada's small population.

Canada also has a strong tradition of amateur astronomy. In 1868, a small group of Toronto astronomy enthusiasts met and formed the Toronto Astronomical Society, which existed in a sporadic fashion until it was formally incorporated in 1890. In 1903, the group received approval to change its name to the Royal Astronomical Society of Canada. Soon, the group expanded beyond Toronto. Today, nearly 5000 people belong to the society, which has centres or local clubs that meet regularly to promote astronomy in 28 Canadian cities. The RASC is famous for its annual Observers Handbook that has become an indispensable reference for amateur astronomers.

Some of the world's best-known popularizers of astronomy hail from Canada. Helen Sawyer Hogg, one of Canada's top professional astronomers in the middle of the 20th century, was known as a writer of books and articles on astronomy aimed at the general public. Today Jack Newton of Osoyoos, BC, is famous for his astrophotography, and Terence Dickinson, who lives near Yarker, Ontario, wrote *Nightwatch* and other popular guides for beginning astronomers and is the editor of *SkyNews*, Canada's astronomy magazine. David Levy, a Montréaler now living in Arizona, became one of the world's top comet hunters and has written many books and articles on astronomy.

Today, telescopes used by both professional and amateur astronomers provide sharper and deeper views of the heavens thanks to advanced computing equipment, some of it developed by Canadians. So even for people who are staying on the ground, the exploration is becoming more exciting than ever.

But many people's urge to physically go into space remains, no matter how much they can see from the ground with telescopes. Fewer than 500 people have gone into space, including the eight Canadian astronauts. But the new millennium brought with it new hope for those who want to fly beyond the Earth's atmosphere without becoming professional astronauts. In April 2001, the first space tourist, American entrepreneur Dennis Tito, flew on board a Soyuz spacecraft to the ISS and spent a week on board the station. Since then, two others have followed. But they had to pay a reported $20 million each, go through rigorous training and pass a demanding medical exam to make their flights, demonstrating that the day when regular people can enter orbit is still some time off.

Three years after Tito's flight, a more promising avenue opened for people waiting for their ticket to space. The first private space vehicle, *SpaceShipOne*, piloted by Mike Melvill, made a suborbital flight 100 kilometres above the Mojave Desert on June 21, 2004. That fall, *SpaceShipOne* won the X Prize competition by becoming the first privately financed craft to carry its pilot and the equivalent weight of two passengers on a suborbital flight to an altitude of 100 kilometres, the generally accepted boundary of space, and back safely within two weeks. The team that built *SpaceShipOne* won the $10 million prize, opening the door to private space travel. Already, British entrepreneur Sir Richard Branson is working with Bert Rutan's Scaled Composites, the builders of this historic vehicle, to offer regular suborbital flights to paying passengers.

The X Prize competition, which began in 1996 and was modelled on the prize that inspired Charles Lindbergh to make the first London-to-Paris flight in 1927, drew 26 competitors from seven nations. Two of the leading competitors came from Canada.

The *Canadian Arrow* team was building a rocket similar to the German V-2 rocket of World War II. They had successfully tested engines and their crew cabin and were training pilots when the X Prize was won in 2004. Now the team has joined with an American entrepreneur to pursue their goal of building the *Canadian Arrow* and an even larger craft to carry passengers on suborbital flights.

Another Canadian team, the *Da Vinci Project*, came close to launching its space vehicle from Kindersley, Saskatchewan, in the fall of 2004, but the launch didn't come off because of a missing part and the expiry of launch permits. The group built a balloon that would carry its rocket to altitude, from which the rocket, powered by a Wildfire Mark VI engine, would fly to the edge of space. Although the flight never came off, the *Da Vinci Project* may take part in future events powered by the X Prize foundation.

Although the Canadian entrants failed to win the X Prize, their efforts mark early attempts in a race that is just beginning—to build a private space tourism industry. There will be some false starts, but these efforts provide hope for those who want to glimpse space and experience weightlessness but can't qualify for the elite ranks of professional astronauts.

Other Canadians are working to advance the dream of space travel through groups such as the Canadian Space Society and the Mars Society, whose international efforts supporting travel to Mars include a simulated Mars base on Devon Island in Canada's Arctic, where engineers and scientists are working on challenges that will face future Mars explorers.

AFTERWORD

A Heritage of Exploration

THE WORK OF THE ENTRANTS IN THE X PRIZE COMPETITION FITS IN well with the history of Canadian space efforts. Some Canadian space pioneers, such as our astronauts and the scientists and engineers who continue to make their flights possible, have worked inside Canada's space program. Others, such Owen Maynard and Jim Chamberlin, made their marks in the history of space exploration when they lost their jobs in Canada and helped NASA put astronauts on the Moon. Still others, like Gerry Bull, made history by aiming high but ultimately falling short. Many others built satellites or used telescopes in anonymity.

Robert Thirsk, one of Canada's astronauts, says Canadians have always shown a strong drive to explore. At a history conference in Edmonton in 2005, Thirsk told the story of a kayak journey he took retracing the footsteps of Sir Alexander Mackenzie, who in 1793 became the first European explorer to cross North America by land to reach the Pacific Ocean. Mackenzie's trip took him across the Rocky Mountains and the upper reaches of the Fraser River before he reached the Pacific near Bella Coola, BC. Thirsk said his studies convinced him that Mackenzie possessed many attributes needed by explorers, including determination and a number of leadership qualities. If Mackenzie had lived today rather than in the 18th century, he would make an excellent astronaut. Thirsk and his fellow astronauts required similar determination to deal with the setbacks they met on the road to space. He said Canadians still have much to do exploring the frontiers of science, technology, medicine and space, among others. "Exploration," according to Thirsk, "is a Canadian core competency." In other words, Canadians are born to explore.

Acronyms

CAE Industries – Canadian Aviation Electronics

CAPCOM – Capsule Communicator

CARDE – Canadian Armament Research and Development Establishment

CRC – Communications Research Centre

CSA – Canadian Space Agency

CTA – Canadian Target Assembly

CTS – Communications Technology Satellite, later known as *Hermes*

DRB – Defence Research Board

DRTE – Defence Research Telecommunications Establishment, a branch of DRB

EOR – Earth Orbit Rendezvous

ESA – European Space Agency

EVA – Extravehicular Activity, or spacewalk

FUSE – Far Ultraviolet Spectroscopic Explorer satellite

HARP – High Altitude Research Program

ICBM – Intercontinental Ballistic Missile

IGY – International Geophysical Year, 1957–58

IML – International Microgravity Laboratory

ISIS – International Satellites for Ionospheric Studies

ISS – International Space Station

LM – Lunar Module

LOR – Lunar Orbit Rendezvous

MBS – Mobile Base System, part of the MSS

MOST – Microvariability and Oscillations of STars, Canada's first space telescope

MSS – Mobile Servicing System

NASA – National Aeronautics and Space Administration

NEOSSat – Near Earth Object Surveillance Satellite

NRC – National Research Council of Canada

RCA – Radio Corporation of America

RCAF – Royal Canadian Air Force

RMS – Remote Manipulator System, the shuttle Canadarm

SRC – Space Research Corporation

SSRMS – Space Station Remote Manipulator System, or Canadarm2

STEM – Storable Tubular Extendible Member

STS – Space Transportation System (the space shuttle)

SVS – Space Vision System

USAF – United States Air Force

UTIAS – University of Toronto Institute of Aerospace Studies

Notes on Sources

Books

Bondar, Barbara, and Roberta Bondar. *On The Shuttle: Eight Days in Space*. Toronto: Maple Tree Press, 1995.

Bondar, Roberta. *Touching the Earth*. Toronto: Key Porter Books, 1994.

Cooper, Henry S.F. *Before Lift-Off: The Making of a Space Shuttle Crew*. Baltimore: The Johns Hopkins University Press, 1987.

Dotto, Lydia. *A Heritage of Excellence: 25 Years at Spar Aerospace Limited*. Toronto: Spar Aerospace Ltd., 1992.

Dotto, Lydia. *Canada in Space*. Toronto: Irwin Publishing, 1987.

Dotto, Lydia. *The Astronauts: Canada's Voyageurs in Space*. Toronto: Stoddart Publishing Co., 1993.

Gainor, Chris. *Arrows to the Moon: Avro's Engineers and the Space Race*. Burlington, Ontario: Apogee Books, 2001.

Grant, Dale. *Wilderness of Mirrors: The Life of Gerald Bull*. Scarborough, Ont.: Prentice-Hall Canada, Inc. 1991.

Harland, David M. and Ralph D. Lorenz. *Space Systems Failures: Disasters and Rescues of Satellites, Rockets and Space Probes*. Chichester, U.K.: Praxis Publishing, 2005.

Harz, Theodore R. and Irvine Paghis. *Spacebound*. Ottawa: Canadian Government Publishing Centre, Supply and Services Canada, 1982.

Jarrell, Richard A. *The Cold Light of Dawn: A History of Canadian Astronomy*. Toronto: University of Toronto Press, 1988.

Jelly, Doris H. *Canada: 25 Years in Space.* Montréal: Poly-science Publications Inc., 1988.

Kirton, John, ed. *Canada, the United States, and Space.* Toronto: Canadian Institute of International Affairs, 1986.

Lowther, Wiliam. *Arms and the Man: Dr. Gerald Bull, Iraq and the Supergun.* Toronto: Seal Books, 1991.

Mayer, Roy. *Scientific Canadian: Invention and Innovation from Canada's National Research Council.* Vancouver: Raincoast Books, 1999.

Murray, Charles and Catherine Bly Cox. *Apollo: The Race to the Moon.* New York: Simon and Shuster, 1989.

Stewart, Greig. *Shutting Down the National Dream.* Scarborough, Ont.: McGraw-Hill Ryerson, 1988.

Sullivan, Walter. *Assault on the Unknown: The International Geophysical Year.* New York: McGraw Hill Book Co., 1961.

Williamson, Ray A. and Claire R. Farrer, eds. *Earth and Sky: Visions of the Cosmos in Native American Folklore.* Albuquerque: University of New Mexico Press, 1992.

Winter, Frank H. *Prelude to the Space Age: The Rocket Societies: 1924–1940.* Washington: Smithsonian Institution Press, 1983.

Articles/Papers

Randy Attwood, "MOST: Canada's First Space Telescope, Part I." *Journal of the Royal Astronomical Society of Canada.* Vol. 96, No. 6 (December 2002), 232–235.

Gainor, Christopher. "The Chapman Report and the Development of Canada's Space Program," *Quest: The History of Spaceflight Quarterly,* vol. 10, no. 4, 2003, 3–19.

Godefroy, Maj. Andrew B. "Defence and Discovery: Science, National Security, and the Origins of the Canadian

Rocket and Space Program 1945–1974." Ph.D. Dissertation, Royal Military College of Canada, 2004.

Shanko, Barry. "The MOST for the Least: Canada's Pioneering Space Telescope," *Spaceflight,* Vol. 48, No. 1 (January 2006), 14–17.

Films
Space Pioneers: A Canadian Story, 1988. Co-directors: Rudi Buttignol and Paul Sobelman.

Websites
Canadian Space Agency
  http://www.space.gc.ca

European Space Agency
  http://www.esa.int

Friends of CRC Association
  http://www.friendsofcrc.ca

MOST satellite
  http://www.astro.ubc.ca/MOST

National Aeronautics and Space Administration
  http://www.nasa.gov

NASA Human Spaceflight
  http://www.spaceflight.nasa.gov

Royal Astronomical Society of Canada
  http://www.rasc.ca/

Chris Gainor

CHRIS GAINOR FIRST DEVELOPED HIS FASCINATION WITH SPACE watching space shots in the 1960s. In 1979, he earned a BA in History. Since then, he has been a journalist and public relations writer, but the call of the stars led him to a Master of Science degree in Space Studies. He is a member of several astronomical and historical societies, including the Royal Astronomical Society of Canada and the Canadian Historical Society, and he is a fellow of the British Interplanetary Society. He has written numerous books and articles on space technology and exploration.

Chris has visited most of NASA's space installations, seen shuttles launched, and even traveled to Russia in 1992 to tour space facilities and witness a launch. These days, he can be found dividing his time between writing and consulting at home in Victoria, BC, and his graduate studies in the Department of History and Classics at the University of Alberta in Edmonton.